Moms
IN THE MILITARY:

RAISING A CHILD
WHILE SERVING IN THE
ARMED FORCES

PATRICIA QAIYYIM

First paperback edition December 2022

Book cover design by Neakail Tolbert
www.neakailtolbert.com

ISBN 979-8-9871951-0-9 (paperback)
ISBN 979-8-9871951-1-6 (e-book)

Published by Patricia Qaiyyim
www.patriciaqaiyyim.com

Congress of Library Cataloging in Publication Data
Control Number: 2022922262

The views expressed in this literary work are those of the author and the women interviewed, and do not necessarily reflect the official policy or position of the United States Government or Armed Forces.

DEDICATION

To Jamal, my husband and partner, whose love and support empowered me throughout my military career, raising our three beautiful children, and this latest journey.

To Jennifer, Jamal II, and Alexis, who endured life as military brats, gave me purpose, and helped me be a better person and a better Airman.

To all the women who served as moms in the military; those before me who helped pave the way and those who serve after me as they continue the privilege, honor, and legacy of serving in the Armed Forces while raising their children.

ACKNOWLEDGEMENTS

I would like to acknowledge and thank God for His grace and blessing. I am thankful for the strength and the courage He has given to me throughout my life. It is my hope and prayer that He continues to guide me as I endeavor to inspire, motivate, and educate through thought-provoking works.

I want to thank my family and friends for their love and support as I worked through the process of writing this book. Their words of encouragement and love can never be repaid. I love you all!

A special thank you to my friend, fellow writer, and mentor, Marcus Griffin. Without your encouragement, I would still be thinking about writing my book instead of finishing my book.

I want to thank Sharon Jenkins, the Master Communicator. Your classes and guidance have truly helped me become a better and more confident writer.

I would also like to acknowledge and thank all the women, named in this book, who shared their story with me. It is your stories that helped make this book possible. I thank each of you for your service to our country while taking on the challenge of motherhood. It is my hope that our stories will encourage other servicewomen who are thinking about service and motherhood and enlighten those who have wondered why we would take on the two unique challenges of Service and Motherhood.

TABLE OF CONTENTS

ACRONYMS/DEFINITIONS

Active Duty: Full time service in the armed forces

Air Force Specialty Code: The alphanumeric code used to identify a specific job in the Air Force

Basic Training: The initial training received by enlisted members upon entry into the Armed Services

BDU: Battle Dress Unform

CDC: Child Development Center; a daycare on military installations

Coast Guardsman: Term used to identify military members of the Coast Guard

Command (Chain of Command): Hierarchical chain of leadership that one follows

Commissioned: Duty as a commissioned officer in the pay grade of O-1 through O-10

Convalescent Leave: Paid leave used as a period to recoup from illness or medical procedures

Deck Officer: A position in the United States Navy, United States Coast Guard, and National Oceanic and Atmospheric Administration (NOAA) Commissioned Officer Corps, tasked with certain duties and responsibilities for the ship

Deploy (Deployment): The movement of military personnel to various locations for possible military action

Dual-Military: Two spouses of active military duty at the same time

Duty Assignment: Permanent location assigned for duty

Enlisted: Duty as an enlisted member of the Armed Forces, in the pay grade of E-1 through E-9

Family Care Plan: A written plan to ensure family members are taken care in the necessary absence of the military member; used by single parents and dual military families

First Sergeant: Dual meaning; In the Army, First Sergeant is the rank position of E-7. In the Air Force, First Sergeant is a duty position held by members in the rank of E-7 through E-9, works with, advises, and reports directly to the unit commander on matters of enlisted morale, welfare, and conduct.

Joint Tour: A tour of duty for members of one branch of service with another branch of service, i.e. joint tour of Coast Guard with Air Force

Judge Advocate: Attorney in Armed Forces

Military Fatigues: Similar to BDU, used prior to the BDU. Known as field uniform or battle uniform

Military Brat: Child whose parent(s) are in the Armed Forces

Military Installation (Base) (Post) (Camp): Locations (installations) around the world where members of the Armed Forces are stationed

Military Occupational Specialty: Numeric code used to identify a specific job in the Army or Marine Corps

Orders: Military instructions; can be written or oral. Military orders are used to assign members to military installations.

Overseas Control Date: The date you returned from overseas duty; can be used to determine next overseas date

PCS: Permanent Change of Station, moving from one permanent location to another

Pickle: A term used to describe Air Force Airmen in the early stage of basic training; indicating newbies moving from civilian clothing (when they are called rainbows) to the Air Force olive drab Fatigue uniform

Rack: Military term used to denote bed or bunk

Rate: Military rank system of enlisted Navy sailor

Remote (Unaccompanied) Assignment: An assignment to a location without family members, not a deployment, usually twelve months or longer

Reserve (Reservist): Military Reserve Forces; typically, part-time in supplement of active-duty service

Service Commitment: Period of time to complete your commission or enlistment in the Armed Forces

TDY/TAD: Temporary Duty/Temporary Additional Duty; Temporary period of time away from your permanent duty location, usually for training, conferences, medical treatment, etc

Technical Training: Training following basic training in which enlisted members learn the skills needed to work in their assigned specialty

TIPS FOR MOMS

- Find your "village" wherever you go. Having a friend (or friends) you can count on makes things easier when you have no immediate family nearby. Having someone you can talk to, vent to, swap daycare with, and laugh with makes all the difference.

- When moving to a new location, get the kids involved before you get there. Learn about the new location, find things that interest every member of the family, and make plans to visit those places. This will give the kids something to look forward to and make the move less daunting.

- Start some family rituals that your kids can grow up with; having these rituals can be helpful in creating some normalcy when you move to a new location.

- If you are going to be gone for weeks or months, and your children are small, try printing out a coloring puzzle for each kid. They can color one piece each day; when the puzzle is all colored, mom should be home. Make sure and add a couple of days when going on deployments as return dates can and often do change.

- Set up times and dates for Facetiming or calls; this can help you kids feel your presence and have some normalcy during your absence.

- Let your spouse help you as much as possible. Trust that even if they may not do it the way you would, they will get it done.

- Let your kids take some responsibility as they grow. This does two things: helps instill independence/work ethic and teaches them life skills while giving you a break on some household chores.

- Make as much time as possible for your kids to get to know their extended families. This will give grandparents, aunts, uncles, and cousins time to enjoy your children and vice versa.

- Kids tend to feel what we feel; when we are happy, they can sense it, and when we are anxious, they can sense that as well. Try and see the silver lining of every situation and share it with your kids.

A CALL TO SERVICE
(SERVICE AND MOTHERHOOD)

For me, the call to service came from within.
To serve my country alongside other women and men.

And with pride and honor, I served without fear.
Wearing my uniform, day after day and year after year.

Before too long, a tiny someone entered my world.
And while I gave her life, she sent mine into a swirl.

For the first time, my journey truly affected another.
I asked myself, do I have the strength to be both Airman and mother?

I looked deep inside and knew without doubt that I could do both.
So, I gathered my strength and readied myself for unfathomable growth.

I loved serving my country, and the places it took me.
Figuratively and literally, and I knew this was where I was supposed to be.

Soon one became three; for my husband and I, this tripled our joy.
But also, the weight of wearing my uniform and raising two girls and a boy.

The decision to serve, I made without hesitation.
Motherhood, I tackled with the same foundation.

Today, I reflect on my service and am thankful for
the opportunity the Air Force did provide.
I am a veteran and a mother, and I am proud to
say I did both with honor and pride.

- Patricia Qaiyyim

PREFACE

Since 1948 women, but not mothers, could serve as full Armed Forces members. It wasn't until 1975 that, without a special waiver, women would have the choice to remain on active duty while raising their children. Almost fifty years later, women continue to prove that motherhood is not a barrier to military service.

Today, almost fifty years later, women are still being asked why they chose service and motherhood. Of course, the answer is, "why not."

The women I interviewed for this book have served or are serving their country and raising their children in the Army, Air Force, Navy, Marines, or Coast Guard. Within these pages, we share our perspective and insight into why a woman would choose the obligation of service while tackling the responsibilities of motherhood. We are proof, fifty years later, that women are capable and ready for service and motherhood.

FOREWORD

Like most children, my mom was the central figure in my early life. I was in awe of my mother. She was amazing. I grew up watching her do the impossible. She was on active duty in the Air Force. She raised three children. She went back to school. She switched career fields. She studied for promotions. She decorated lockers. She put patches on our pants and wrote notes on our napkins. She came to pep rallies and soccer games. She was a mom to us and a mom to her troops. And she made it look easy.

When you go to school on a military base, having an active-duty dad is expected; it's almost a given. Having an active-duty mom is special. "Your mom's in the Air Force, too?" friends would ask with wide eyes. "Who does the mom stuff at your house?" they wanted to know. "My mom..!" I would reply to their raised eyebrows. This was the only life I ever knew, and I was confused as to why they would even ask such a question. Of course, my mom is the mom at my house.

Military Moms are in a club that is, thankfully, becoming less and less exclusive. It isn't an easy road, but they keep going. This book is an answer to anyone who says women can't-or shouldn't-be a mom and wear the uniform. They can. And if they want to, they should. Those women raise children like me, who are proud of their mother's military service.

Serving in the Armed Forces is an honorable calling. It's a grave responsibility. It requires sacrifice. It commands respect. And motherhood is no different. Read this book and look inside the lives of the women who dared to answer both of these callings-who dared to forge these two paths into one.

Anytime a woman does something, she makes it a little easier for the women who come after her. These women changed the course of history by choosing service and motherhood. Read their stories: stories about parenting, duty, and camaraderie. Like all good stories, you will find that you learn something about the world, and you learn something about yourself as well. Military moms, working moms, single moms, stay-at-home moms-this book is for all moms-and the people who love them.

I'm all grown up now, and I am still in awe of my mother. She is still amazing. She continues to inspire me as she opens new doors and chases new dreams.

-Jennifer Vamper

CHAPTER 1:
AN INTRODUCTION

**Instead of seeing my strengths
and how I could contribute,
he only saw my gender.[1]
-Maj Mary Hegar**

No matter where you look in our world, you are bound to see a working mom. Since the early 1970s, working moms have been included in the Armed Forces. Today, many of us know a mom who works outside the home. It is no longer a rarity, or something done as a necessity; for some, working while raising children can be a choice. A choice that many of us have made. A job can provide many things, including monetary gain, giving back to society, setting a positive example for young girls everywhere, or taking a God-given talent and making the world a better place. Whatever the reason, women work outside the home and succeed in every conceivable field, including all branches of the Armed Forces.

Women served in the military when it suited the military. Still, it was not until 1948 that women were allowed to serve as full Armed Forces members when President Truman signed the Women's Armed Services Integration Act[2]. Then, while we could serve in uniform, the choice to serve while raising a child was not always given to women serving in the Armed Forces. Between 1951 and 1975, pregnancy meant automatic discharge from service for any woman who became pregnant, regardless of her rank or time in service; this included women who adopted or, through marriage, became a stepmother[3]. This executive order also excluded women with children from entering service. In 1971, the military instituted waivers of discharge for pregnancy; in fact, my sister was a part of that first wave of women service members who were also mothers when she entered the Navy with a young son. Finally, in May 1975, the Department of Defense (DoD) ordered the Armed Forces to do away with involuntary separation for pregnancy[4].

Even with this new order, while women were eager to serve while raising a child, some branches were reluctant to discontinue the mandatory discharge. As late as 1976, some Armed Forces branches were still separating pregnant servicewomen, whether or not they wanted to separate. But that same year, the Second Circuit Court held that the mandatory separation upon pregnancy violated a servicewoman's Fifth Amendment[5].

For me, what really highlighted the seriousness and heartache of this mandatory separation policy was an episode of one of my favorite childhood shows, M*A*S*H. In episode nineteen of season six, Major Houlihan is a career Army woman (following in her father's footsteps). She is suddenly faced with the possibility of pregnancy and the pain of knowing that one thing she wanted would mean giving up the one thing she had[6] (her career in the military). Of course, it was just a television show, but I can now imagine and empathize with the many women who faced such a dilemma. And today, there are still women painfully remembering their separation and the end of their military career with benefits or compensation. In fact, earlier this year, this subject was in the news when it was announced that Congress is addressing this very issue and has plans to amend the upcoming 2023 defense policy bill, declaring it was a mistake to force women to leave the military during pregnancy[7].

In 2020, according to the 2020 Demographics Profile of the Military Community, the active-duty Armed Forces (Air Force, Army, Navy, and Marine Corp) was comprised of 17.2% women, numbered at 229,933. Of the 229,933 servicewomen on active duty, approximately 30% had children[8]. According to the same report, the number of moms serving in the Armed Forces Reserve and Guard totaled 58,718 (about 34%). Looking at our Coast Guard branch of the Armed Forces, of the approximately 40,000 active-duty members, 13% are women[9]. While I am not sure what percentage of active-duty Coast Guard women are moms, you can be sure that there are women in the coast guard serving our country and raising their children.

Many women are primary caregivers to their children and being a mom in any situation is unique and requires work, dedication, and love. Moms in the military are not better than other moms but have answered a call to a career and set of circumstances just as unique as motherhood. Serving in the Armed Forces is genuinely a "way of life." Much different than life in the civilian sector. So, what makes a military mom so distinct? The same thing that makes a career in the Armed Forces so unique.

Life in the Armed Forces involves being on duty twenty-four hours a day, seven days a week. It includes living in unfamiliar or unwanted locations, hundreds and even thousands of miles away from your family and/or support system. Even the subtle freedoms of choosing what to wear and how to wear it are a sacrifice. In the Armed Forces, everyone is told what to wear, how to wear it, where to live, where to work, when your family can join you, and even when they cannot. For moms in the military, that can mean leaving their children for weeks and months at a time.

This may seem an impossible order to follow for those not serving in uniform. When speaking with non-military moms, they would tell me that they could never leave their children for extended periods. But moms in the military, myself included, have come to consider and accept separation, not as something we certainly looked forward to doing, but a necessary part of sacrifice to service.

Serving our country and raising our children presents unique challenges that most mothers in the civilian sector may not have to consider or experience. I am glad to see so many taking the challenge, serving our country, and raising their children; and meeting these challenges in their own ways and on their own terms.

There is another mom who, although she did not take the oath of enlistment or commission, still serves as a part of the Armed Forces. She does not wear the uniform, but she sacrifices just the same. She is the military spouse whose experiences closely match that of moms in the military. She, like us, lives the unique lifestyle that the military offers. She shares our uniqueness as her husband serves his country. Maybe even more so because she is often forced, as a wife, to be a single parent while her spouse is deployed, on Temporary Duty (TDY), or serving in remote locations where spouses and families are not allowed. She is living a lifestyle that, only through marriage, she has agreed to. Living a military lifestyle without the benefits and pay that service members enjoy. She is often living away from the support system that she grew up with, often moving without the promise of a job, and raising her children while facing the challenges of life in the Armed Forces. One such challenge for these women is finding a career that works well with military life, or for some, putting a career off in support of her husband's military career.

During most of my twenty-plus years on active-duty service, I was often asked, as a mother, how or why I chose this lifestyle for myself and my children. These questions came from family, friends, my co-workers in uniform, and civilians I worked with or met throughout my career. Over time, I learned that I wasn't the only servicewoman questioned. And honestly, there were times, while still on active duty and feeling the pressure of military life, that I wondered why I, or other servicewomen, chose to stay in the Armed Forces while raising their children. In the end, although not always easy, I knew I wanted to continue my service, fulfilling my dream, and showing my children that life in the military was possible.

As a first sergeant, I was required to review Family Care Plans periodically. These plans are required for single and dual military parents to ensure the necessary steps are in-place to provide for the care of children when the parents are called away on short notice and/or to deploy. While reviewing Family Care Plans, I sometimes chatted with my fellow servicewomen and learned that while we all make the same decision, it is not for the same reasons, nor do we tackle the challenges in the same way. I discovered more about military moms and the uniqueness of our choice of service and motherhood and gained more insight into why we chose life in the military and how we manage it.

Even after retirement, I was still asked why some women continue to serve in uniform and why other servicewomen decide to honorably separate and raise their children. I couldn't answer for other women, but I do know that both decisions are honorable, and each servicewoman should be appreciated and thanked for her service to our country.

To satisfy my curiosity and write this book, I interviewed past and current servicewomen to learn more about their decisions. To get as broad a perspective as possible, I interviewed women in the Army, Navy, Marines, Air Force, and Coast Guard. These women agreed to be interviewed and share their stories so that I could learn more about servicewomen raising their children. I also interviewed military spouses facing some of the same challenges as moms in uniform in my quest for answers. This included women I have known for a lifetime, women I have known for a few years, and women whom I met only through my desire to speak with other women who are also serving their country as moms in the military. I discovered, that like me, many women joined and became mothers; for others, motherhood would come first, taking on the challenge of serving as a mom right from the start of their military careers.

In this book, I share the thoughts and decisions some moms make about simultaneously serving their country and raising children. Not just my thoughts but those of other women who are also moms in the military. Through this book, I hope to give moms in the military a voice and a chance to share their story and perspective of why they do what they do and how they manage the obligation to service and the responsibilities of motherhood.

17

I hope this book is a source of encouragement for all military moms raising their children while serving their country. Finally, I hope this book will be a source of information and understanding to those who want to know how or why women choose service to country and motherhood. And as I finished this book, I realized that I also hope it would speak to and encourage all working moms, those inside and outside the Armed Forces.

CHAPTER 2:
MY OWN JOURNEY

The most effective way to do it, is to do it.[10]

-Amelia Earhart

When I joined the military, I didn't consider motherhood. I was focused on doing what I wanted, which meant serving in the United States Air Force. While motherhood would come shortly after joining the Air Force, I hadn't considered having children nor thought about the challenges motherhood would bring to my military career.

I decided early in life that I wanted to be in the military. My desire came from my time in the Junior Reserve Officer Training Corps (JROTC) and my love of the television series *M*A*S*H*. As any of my siblings will tell you, I was drawn to the show from the start. While the show was based on being in the field during the Korean War, it also focused on the members of an Army medical unit and their striving to do their part. For me, it provided an inside look at life in the Army; it touched on serious issues but also showed the lighter side of life in the military, and in these characters, I saw support for one another, camaraderie, a willingness to speak up and the ability to find the good and the funny, even during a war. This included those who willingly served and those who were drafted to serve. They were able to put their differences aside and get the mission done. I knew I wanted to be a part of something like that. My time in JROTC showed me that the Air Force was the branch for me. I got to experience a little of what life in the Armed Forces might be like; I even got to travel to military installations and see servicemen and women in action. I even tried to enroll in a military high school but was denied admission because they were an all-boys school at the time and suing for entry was unheard of.

Before even joining the military, I knew I wanted to make a career out of service. My oldest sister and brother had both served in the Navy; I had cousins and uncles who also spent time in the military. What I saw in their actions were their desire to do more, to be more, or maybe their sense of patriotism; and this made me want to join the ranks of the military…and perhaps it was *M*A*S*H* or my time in JROTC that reinforced my patriotic spirit. Either way, I found that I wanted to be in the military just to be in the Armed Forces. I simply wanted to serve and experience all it had to offer.

As ready as I was to join the Air Force, my parents were not, even after having two other children serve in the Navy, or maybe because of it. So, with my parents' urging, I opted for college right out of high school; I was more persuaded by the fact that I was only seventeen, and they didn't want to sign the paperwork. I couldn't blame them; they thought I should go to college, something neither of them had the opportunity to do. Still, I knew going to college was more for them than for me, and I decided to enlist after two years. When I told my parents that I planned to discontinue school and join the military, they were not surprised. And I was able to complete my degree while serving, something they wanted for me all along.

I began my service in 1985 (I find it hard to believe it was that long ago) when I joined the United States Air Force and the ranks of the many women who came before me. When I enlisted, I planned to make the Air Force my career and looked forward to the great adventures I thought being in the Air Force would bring. I was extremely excited about my enlistment and was ready to go. I even enjoyed basic training, or at least most of it. Do not get me wrong, it was not exactly fun (and I never knew people could yell so much), but it was the beginning of a journey that I was so excited to start. Those weeks in San Antonio, Texas, were my first introduction to the "real" military and seemed to go by more quickly than I thought they would. For me, the days went by fast and the nights even faster. The women I met, and our bonding, helped us get through basic training and was an excellent introduction to the camaraderie that I thought the military would provide.

While basic training went by fast, I cannot say the same about my job training. With basic training, there was very little freedom, and everything was laid out for me. Tech school required more work on my part; and my training for a job in maintenance was more challenging than I expected. Add to that, I was twenty and did not expect such restrictions to be imposed. But like basic training, my fellow Airmen and I bonded and worked hard to ensure we all made it through training together. Finally, after four months of training, I was ready to be an Aerospace Ground Equipment Apprentice (mechanic). I knew I was ready and looking forward to a change, and the Air Force provided that change. My first assignment was to Zaragoza Air Base, a small military installation in central Spain. I don't remember at all being afraid of such a major life change. Probably because I felt my whole life had prepared me for life in the military. I was excited, much to the dismay of my mother, who thought it her duty, to call my commander, much to my embarrassment, and tell him that I was not ready to go to Europe. Luckily for me, commanders don't listen to moms, and before I knew it, I was headed to Spain.

I was excited to discover what life would be like once I got to Spain. But I soon realized it was the best first assignment I could have hoped for. Being in a foreign country was ideal for the camaraderie I'd imagined I would experience while watching episodes of *M*A*S*H*. Coming from a small Midwest city, I felt some culture shock, but in a good way. Getting the opportunity to live off base really enhanced my whole experience. So, while I thought before I joined the Air Force that it would be for life, I now knew that my joining the Air Force was the ideal career choice for me. I really felt like my whole life had prepared me for military service. Everything from having to do chores, to frequently moving, to having a large group of people (my family) that I could depend on and could depend on me, to being part of something (again, my family) more significant than me. All these things my parents instilled in me helped as a young adult.

Leaving my family and the familiarity of the United States for the first time as I headed to Europe would be the biggest challenge of my life up to that point. I was excited and nervous at the same time but looked forward to the adventure. I was ready to take on the world and all it had to offer. After a couple of years in the Air Force, I still loved service in the military and was excited to continue my journey.

While Spain was my first challenge, it would not my biggest one. Just a couple of years into my career, I faced my greatest challenge: motherhood. Any nervousness I might have felt when leaving my home for a foreign country was nothing compared to the anxiety that came with learning I was pregnant. And I wondered, for a moment, how I would manage it. While letting everything sink in, my supervisor and doctor explained that I could choose to separate and raise my child as a civilian. To my surprise, my first sergeant told me he thought it would be best if I separated to raise my child. I was surprised and frustrated by his outdated attitude. But once I let everything sink in and thought about what I wanted, I knew I wanted to continue my Air Force journey and that I was up to the challenge of service and motherhood. At that time, I could not relate to the forced separation of previous decades, and looking back, I was thankful I could choose to stay or separate rather than be forced to separate.

Even though it had been years since women were allowed to serve while pregnant, I remember being pregnant, as a maintenance troop, and not having a maternity uniform to wear. At that time, the Air Force did not have maternity uniforms other than   blues. Once my pregnancy prevented me from wearing my fatigues (which would later become battle dress uniforms), I had to wear sweats under my coveralls, which was crazy and hot in July and August. I also remember being in a male-dominated specialty, and while my pregnancy did not in any way alter my performance, it did affect my appraisal, something that shocked me. I truly felt that my leadership was holding my pregnancy against me.

When my second child came along, I was prepared to continue service and raise two children. My leadership, especially my commander, was more understanding and supportive of my role and my ability to handle my job and my family without my performance suffering. My confidence in my ability to do both was greater than ever. I also knew that my husband was right alongside me and would continue to support me in raising our family and my career. Really, I thought, two kids and two parents: evenly matched. Coming from a large family, I thought, "How much harder can it be?" After all, I watched my mother raise a dozen kids. It was a little more complicated than I thought but manageable. Having two kids with a six-year age difference meant two drop-offs in the morning, two pickups in the afternoon, and two children to feed, bathe, clothe, and even entertain. At times, it felt like double everything. But soon enough, we got the hang of it.

We barely had time to adapt to our new situation, but felt we had things running smoothly, when we discovered we were pregnant again. I was near the end of my special-duty assignment, and my training for a new job was right around the corner. This would mean a move with two children under the age of two. In fact, before our move, I went TDY to complete training to become a paralegal when my youngest was only a few months old. Not the first time I had to leave my children, but the first time I would go TDY when one of my children was so young.

While having my third child started out with the challenge of attending training for a new job and PCS'ing (Permanent Change of Duty Station), I found that three children were not much more demanding than two. My oldest was six years older than her younger brother and a little more independent. The fact that my son was not even a year older than his younger sister didn't add too much pressure, or maybe my mom was right in that as you gain more experience as a mom, it just gets easier. So, shortly after my third child was born, we moved again, with our own three kids, my young niece, and a new job. Once we settled in, I felt I was hitting my groove in both my service and motherhood. My oldest child would benefit from this move. This would be the first time she would attend school off-base. This meant developing friendships with non-military kids. I am sure her experiences up until this point have prepared her to meet the new challenge, but it was eye-opening for her. It was a big surprise for her was when she realized that other kids lived in the same town as their aunts, uncles, cousins, and grandparents. Until now, I don't think she considered how kids outside the military lived. While she had visited her own family and saw this, I don't think it was until this point that she realized that there was a lifestyle, unlike the military.

After three years, it was time for another move. This time, we were headed to Japan. I saw this as a new opportunity and adventure and did all I could to share that sentiment with my children. We got ready for the move by learning about the culture and language while thinking of things we could do while living in Japan. This would also be our longest in-place assignment. We requested and got approved for another three years in Japan so that my husband to take advantage of a career-broadening opportunity.

While we missed our family, our time in Japan was incredible. We went home at the three-year point to visit family, and several members from both our families came to visit us throughout our time in Japan. These were experiences that our family would not have had if not in Japan, and I know we all grew from our time there.

Six years into our assignment in Japan, my husband retired, and PCS'd to California; that left the kids and me in Japan, where we would stay for another year until our oldest graduated from high school. By this time, I was a senior NCO, which meant more responsibility at work. I remember heading to the States for a TDY after my husband had left, which meant my seventeen-year-old daughter was "in charge" of her younger siblings for about ten days. We had friends there to check in but leaving them alone in a foreign country was hard. But they managed, and I think they enjoyed that time and the freedom. My husband's new job in California was quite challenging for us all. Adapting to life without his help was an adjustment. Also, I became a first sergeant soon after my husband left, which meant my work schedule was not always the same, and it required a lot of my time; I had to be available twenty-four hours a day, which meant responding on evenings and weekends as well. This also meant that the children had to be a bit more self-reliant. It was a time of growth for all of us, which left us all the better for the experience. As for my children, they developed a greater sense of independence, and for me, being a first sergeant was the best job I could have had during my Air Force career.

What I learned most from my decision to take on motherhood in the military is that life as a mom is not a perfect way of life, and sometimes it is not easy. And the sacrifice to serve my country was not only my sacrifice but my children's as well. While they did not have a say in the matter, the benefits outweighed the costs for all of us. Do my children agree? Well, that's for another book.

Throughout my career, I was sometimes asked how I did it...serving my country and raising my children, I was often asked why I chose to do both. This came from strangers, acquaintances, friends, family, and even my fellow service members. Sometimes, the questions were: Why not get out of the Air Force and raise your kids? Why raise your kids so far from your extended family? Why be separated from your children for weeks or months on end? Why put yourself in harm's way, uproot your children every three to four years or make your children pay the price for your service?

For me, the question was not "why" but "why not." Why not have a career in the military and raise my children? Why not experience the uniqueness of service in the Armed Forces? Why not raise my children while wearing the uniform and give them unique experiences that would develop, shape, and allow them to see the world and experience diverse cultures up close and personally? Why not take on the experience of a one-of-a-kind career and lifestyle for my family and me?

When I look back over my life and career, there are some things that I would do differently, but serving my country and raising my kids simultaneously is not one of them. I remember being on cloud nine when I donned my first set of olive drab fatigues in basic training and moving from a "rainbow" (wearing civilian clothing upon arriving at basic training) to a "pickle" (donning my first green fatigue uniform). I truly enjoyed my years in uniform, all things considered.

During my service, there were many periods of excitement and joy but also periods of frustration and anxiety. Looking back, if I were to ask myself,

"Was my time in the Air Force one long season of *M*A*S*H*?" Absolutely not. Sometimes I wanted to scream, cry, and even walk away from life in a uniform. But those times were short-lived. And, while life in the military may seem like constant change, I found a sense of stability in my service; that sense of stability, and my faith, got me through every day. Life in the military can be challenging because it requires so much of you, but for me, it was what I wanted, and I gained discipline, strength, my own voice, and camaraderie during my many years in the Air Force. I am proud of my service and glad it was my chosen path.

Even after I retired from the Air Force, I was often asked about my decision. I was still asked why I chose service while raising my children. Then and now, many years after my service, I want to help others understand why some servicewomen choose service and motherhood. My desire to write has been with me for a number of years, and I could not think of a better way to kick off life as a writer than with this book.

Because I knew that women who chose service and motherhood were still being asked questions about their decision to raise their children while in the Armed Forces, I decided to follow through with my desire to author this book. To provide insight into why women choose to raise their children while serving their country.

In the years since my retirement, I have seen the growth and changes in the Armed Forces regarding women in general and moms specifically. Even if it seems like the changes are coming slowly, I have seen them, and our nation's leaders continue to make positive changes, impacting women in uniform. Especially considering that for many years, there were certain military jobs or areas that did not expect or want women to serve in; or that women could not serve while raising children. I know that the Armed Forces still has room for improvement regarding women in service (those choosing service and especially those choosing service and motherhood), but I can see continued change.

CHAPTER 3:
THE DECISION TO CHOOSE SERVICE AND MOTHERHOOD

When you hold your baby in your arms the first time, and you think of all the things you can say and do to influence him, it's a tremendous responsibility. What you do with him can influence not only him, but everyone he meets and not for a day or a month or a year but for time and eternity.[11]

-Rose Kennedy

There's an old saying, "If the military wanted you to have a family, they would have issued you one." But the military has come a long way. Having a family while serving in the military is commonplace and even the "norm." Choosing to continue service after pregnancy can be a simple decision for some women and a tough decision for others. Thankfully, today it is a choice each woman gets to make for herself and her family, and pregnancy no longer means the end of a military career.

I know parenting is a full-time job, whether you work outside the home or not. For each parent, it is unique, demanding, requires sacrifice, and allows no time off. Even when away from their children, moms are still moms. Being a mom doesn't stop when you walk out the door and head to work. It is, from my point of view, one of the most unique things a woman can do. Whether a woman works outside the home by choice or necessity while her children are growing up, being a mom is challenging and extremely rewarding. Today, moms working outside the home while raising their children include women who choose a career in the Armed Forces, which is also a unique decision.

Serving in the Armed Forces is like no other career. Like motherhood, it is unique, demanding, and requires sacrifice. Serving in uniform means that some of your career decisions are out of your hands, from what you do, to where you live, to how long you can serve. You are on duty twenty-four hours a day and have the distinct and demanding obligation of protecting our nation. You must comply with many orders, which can weigh on you; some are easy, and some are not so easy. It requires much of you and your family. For women who choose service and motherhood, these two responsibilities are constant but not an either-or option. They can be and are done in every branch of our Armed Forces. I am lucky to have served when motherhood was not a mandatory separation. For me, that would have been a heartache. I had always wanted to serve, and I am unsure what path I would have chosen if I did not serve in the Air Force.

In researching this book, I felt so proud of the contributors. Their words and decisions to serve humbled me. I am amazed at how each of them managed daily life in the Armed Forces. Even though we all chose different services and had different reasons for our choices, I felt connected to them in our "sisterhood of service."

While the military is not a sorority, I do feel that joining the military and serving with other women, and especially mothers, invoked in me what I imagine sorority sisters experience. Even years after my retirement, I feel a link to these women that only having served in the Armed Forces can bring. In fact, I don't think you ever really leave the military when you serve for decades. Instead, you just transition from 'active' to 'retired' status. I feel such a connection to military moms that I was more determined than ever to share our stories in hopes that it might help or encourage other women in the same situation or who are contemplating the option of motherhood and service.

The choice was not to put service to our country ahead of motherhood but instead to have a family and serve one's country. Is the decision to continue service while tackling motherhood the right one? It may not be the right choice for every servicewoman, but it was the right choice for many servicewomen, including myself. It was a decision that many of us would do all over again if we had to. That option looks different for each of us; we chose different branches and how we serve. Some of us are on active duty, others select the Reserves or Guard, and some choose to support their spouses as they make the military a career.

Choosing motherhood and service is not done for one reason but for many reasons. We make the same decision for ourselves and our families, but the rationale is as unique as each of us. We chose this path for our own reasons, and below, we share our thoughts on our decision to select service and motherhood.

> *When I learned I was pregnant, my mind was reeling with questions. Could I handle it? Was I ready? What would my family think? Was I sure this was what I wanted for myself? But, once the initial shock wore off, I do not believe I seriously considered separating from the Air Force. My thoughts and feelings about the Air Force didn't change once I became pregnant. I still had a strong desire to serve and did not see a reason to compromise my dream of a career in the Air Force. I enjoyed being in the Air Force and knew it was what I wanted to do. Honestly, I knew motherhood would not change my need to have an income, and I could not think of another career I wanted outside the Air Force.*
>
> -Patricia Q., E-7, USAF (Ret)
> First Sergeant and Mother of three

I decided to join the military as a way out of my hometown; I felt it was my only option in the 1970s. Going to college didn't seem like an opportunity for me, and I saw the military as a way for me to have a future. It was a different time, and they didn't expect women to do anything except women things (which didn't include the military). I was a single parent when I enlisted, and it was the best decision for my son and me. Even though we were separated for my training, it was my only option. I also knew serving in the Navy would be a way to go to college.

-Jacqueline G., E-3, USN Veteran
Personnel Specialist and Mother of two

It wasn't even a question because my number one goal when I entered the Air Force was to serve a full twenty years. When I found out I was pregnant, I knew I would continue to serve in the Air Force. Getting out wasn't an option for me. I wanted my daughter to know what I did, that I was really proud to be in the military...and that I could be a mom and be in the military at the same time.

-Anna P., E-5, USAF
Services Specialist and Mother of two

As a military spouse, it was my husband's choice to serve in the military. So, I really didn't have any thoughts about his service. I knew I would be moving, but that was it. I was going to be leaving home for the first time, which was a bit nerve-racking. For me, it was that initial move from home to England for four years. I had never been to another country or away from family and being so far away was constantly on my mind.
And having my sister and her husband stationed in England really helped me adjust. Of course, they were only there for our first year, but it was nice to have them nearby.

-Carol G., USAF Spouse
Teacher and Mother of one

The decision to remain on active duty was easy because I was married to another military member. We both knew what it would entail, that we could be stationed together, so there wasn't an issue with me staying in.

-Pagerine J., O-5, USAF (Ret)
Medical Service Corps and Mother of one

I didn't have a decision to make. I was going to be a mom, go on maternity leave, and go back to work...not a decision for me, at all.

-Leslie L., O-5, USA
Human Resource Officer and Mother of twelve

For me, the choice to continue service wasn't tough at all. I had been in for ten years, and I had a plan. My plan included getting my master's degree and later my commission. I never even thought of getting out, but my husband and I both decided I would get out at the twenty-year mark, which I did.

I think for many outside the military, their perception of the military is that of a wartime environment. That's only a part of what we do, and continuing to serve is more than that.

-Yulanda B., O-4, USAF (Ret)
Personnel Officer and Mother of one

My father was in the Army, and he had been telling me for years to join. And then, one day, I decided to join. It was a way out of my hometown; I couldn't afford college, and the Army was a way for me to get out and do something and make something of myself.

My husband and I had been trying to get pregnant for a while, and I actually found out I was pregnant when I was retraining at a time when I needed to retrain. So, I hid my pregnancy during the first part of my retraining so as to not be sent home as per Army regulations. Luckily, when it came out, my commandant (who was a woman) allowed me to stay in and complete my training.

I had been in for nine or ten years, and fortunately for me, my husband was a civilian by then. So getting out was not an issue for me. Because I had not yet deployed, the thought of deployment never crossed my mind. I just thought I'll have my baby and keep going. Separation was never an option for me. In fact, my supervisor didn't even talk to me about separating or my option to separate.

-Eboni B., E-8, USA (Ret)
Career Counselor and Mother of two

There was hesitation; I knew I wouldn't serve without my daughter, so I had to figure out how to work it out. Even though I had been a military spouse, coming on active duty as a mom, there was a hesitation. My first assignment was overseas, and the biggest concern was childcare for my daughter. We ended up having my husband's cousin come to Germany and help us for a couple of years.

-Patricia H., O-5, USA (Ret)
Judge Advocate and Mother of three

At first, it wasn't a hard decision; I knew about all the benefits of military service, so I just assumed I would continue service. It didn't really hit me until the end of my convalescent leave. I was like, "I can't go back to work," but I did go back to work. My husband and I thought about being close to family. We discussed it, and I didn't want to have to start over, and I would have had to start over. So, after weighing everything, it really was not a hard choice to stay in.

-Toni D., E-8, USAF
Personnel Specialist and Mother of three

My husband joined the Air Force after we were married and already had one child. However, my son was born while my husband was away at training. While I expected that we would move frequently, I did not expect to be and feel alone so much and have to do things on my own because he was away. I would jokingly tell my husband that he was married to the military, and I was the mistress.

-Ebony A., USAF Spouse
Mother of two

My daughter was born after being on active duty for four years. The option to not continue service wasn't really a consideration for me. I knew my child would depend on me and the benefits offered were a way for me to provide for my family. At that time, we lived in Germany, and my husband was no longer on active duty. I was the primary breadwinner while my husband was going to school.

-Kimberly B., E-5, USA Veteran
Logistics Technician and Mother of four

I joined while married with a child. At the time, my husband was not working, and I needed a way to support myself and my family. So I joined the Air Force for myself and my daughter. I wanted to provide for my family, and this was an ideal way for me to do this.

-Jennifer H., E-3, USAF
Intelligence/Cyber Specialist and Mother of one

Having been on active duty for a while, there was no decision; I would stay in. No decision: I love the Air Force, and I love serving. My husband was separating from service, and I wanted to continue my service. My goal is to make CMSgt. My husband was already separating from the military, so it made sense for me to stay in; my husband became a stay-at-home dad for a couple of years.

-Chelsea D., E-7, USAF
Medical Lab Technician and Mother of three

I was already married to an active-duty member when I joined. My husband and I decided it would be good for our family to join.

The separation from my kids for training was difficult because I had never been away from them. I can't lie; it was hard. I remember several times, I was in my rack and would wake up thinking my baby was about to fall out of the bed. So, getting used to being by myself was hard. But when I reflected on our plan, it made it easier to get through the separation.

-Mykeia L., E-3, USN
Logistics Specialist and Mother of five

I grew up around Scott AFB and was very much around the military and worked with military spouses. And my husband and I, knowing what we wanted out of life, decided to consider the military, and after looking into it, we decided it was a good path for us. I was already a mom and realized that the Air Force had so much to offer my family, not just me and my husband, but also my children.

-Kelsey Q., O-2, USAF
Operating Room Nurse and Mother of two

I came in to do my four years and get myself together but ended up staying. Time went on, and nine and a half years later, I applied for a commission. Staying on after I had children wasn't a hard decision; I came in at twenty-three and had been out in the world, and that experience helped me to understand the benefits the military would offer. And I was on active duty for almost ten years before we had children. It helped that my husband had been in the military and was retired, so I knew he would help out when I was working twelve- and fourteen-hour shifts.

-Cynthia S., O-4, USA (Ret)
Nurse and Mother of two

I came into the Coast Guard out of the Coast Guard Academy and started as a Deck Officer when there were only about 10% females serving. After two tours at sea, I applied to law school and became a judge advocate for the Coast Guard. My path to motherhood is a bit different. I had been in for a while, and once I started embarking on the journey of becoming a mom, I realized I had to have a child. It was exciting, and I ended up adopting in Florida as a single parent and ran into an old boyfriend; we reconnected, and he was a doctor and a part of the birthing process, and we ended up getting married. Many folks were asking me if I would get out, but I thought, why would I? I was established in my career, and getting out didn't occur to me. When I started the journey, I was single, and this was my way of supporting myself and my child. My husband and I are not the typical military family; we have more financial resources than some other military members.

-Melissa B., O-8, U.S. Coast Guard
Chief Counsel and Judge Advocate General and Mother of one

Coming out of the Naval Academy, I had a five-year obligation; if you had asked me about my plans in college, I would have said I would do my five years and move on.

Once I got pregnant, my decision was pretty straightforward because I had not completed my five years and didn't want to resign without meeting my obligation. So, I knew I would be on active duty after having my child and see how it would fit in a way I hadn't planned.

-Janell H., O-3, USMC Reserves
Logistics Officer and Mother of one

Right out of college, I joined the Marine Corps and was on active duty for ten years before joining the Reserves. When I joined the Marine Corps, I didn't think about having kids; the Marine Corps was my life. During my first tour, I was so busy and consumed with being in leadership roles and loving what I did. When my husband and I did decide to have children, it was difficult for me to get pregnant. And that struggle was hard. I am a Marine, and when I decided to do something, I did it. We were finally pregnant, and a few weeks after finding out, I was supposed to be attached to an infantry role, which I thought was great. I had to tell my boss, and he was super understanding (and the exception). He was excited for me and told me not to worry. And I was stressing about letting the team down and having to have someone else take my place, but I didn't think about getting out at all.

Fast forward, my child was six weeks old, and I was struggling with the thought of returning to work; but at three months, I was ready to go back to work. Going back to work, I did get some questions from people at work about my returning. As a combat engineer, I was the only female in my unit. Some of my coworkers were older and from a different generation, and they were confused about and questioned my decision to return to work. Looking back, it is easy to realize that they were just asking me dumb questions, but at the time, it did make me think, "Am I doing the right thing?" I asked myself, "Am I a bad parent?" But plenty of folks supported me and ensured I had what I needed. I remember doing P.T. shortly after returning to work, and the general we had at the time told me it was okay to take some time to get back into things and to take it easy and do what you have to do. That was a huge relief to hear him say that. I was very fortunate at work.

-Corey G., O-3, USMC Reserves
Combat Engineer Officer and Mother of two

I was commissioned in 2012, and I have had the opportunity to work with a lot of good officers. And, In 2018, while in Staff Platoon Officer training, I got pregnant with my first child, and was removed from that position due to the health risks/liability, demands, and stress. That was a bit emotional in that someone else made the decision to remove me from training. At the time, I was frustrated but looking back, I am okay with it. I would have regretted it if something had happened because I continued the training.

After finding out I was pregnant with my first child, my husband and I discussed it, and we decided I would stay in.

I love the Marine Corps, and I knew that we would have to make some sacrifices, so we laid it all out while making our decision. I didn't want to have to choose one or the other. I had not seen a lot of women in the Marine Corps doing it and thought I could be an example to other women and leaders on how to adjust to women who want to take that route.

It was a hard decision. We talked about the sacrifice we would have to make as we're both officers and worked really challenging jobs.

When I got back to work, I got a ton of questions from a lot of people; it was almost like it was the expectation for me to get out of the Marines.

-Maria F., O-3, USMC
Training Officer and Mother of two

I had been serving for eleven years before I became pregnant with my first child. It was not a hard decision. I was already halfway through my career, so I never even thought of getting out. My husband and I are both Marines. If I were pregnant, I would have my child and keep going. I didn't think of it as an either-or situation. I just knew I was going to go back to work.

-Jennifer P., E-8, USMC
Administrative Specialist and Mother of two

I entered the Air Force after graduating college in 1999; I met my husband and got married, and after our third move, we had our first child. I don't think it was a hard decision to stay in. I probably had a service commitment, so it wasn't really a choice. I remember us thinking about getting out at some point during that third assignment but at that time, that is also when the economy crashed around 2008, and I remember thinking we would still have to get jobs, so we stayed in.

-Rojan R., O-6, USAF
Program Manager and Mother of three

CHAPTER 4: FAMILY'S REACTION

The question isn't who's going to let me; it's who's going to stop me.[12]

-Ayn Rand

Many people, myself included, can't wait to share the good news. In the Armed Forces, we are often far away from the family we grew up with but still enjoy sharing our accomplishments. Sometimes, it's a promotion, an award, or a new assignment. Sometimes, it's a new addition to the family. The first reaction is often one of joy and sometimes surprise, and sometimes, when you are a woman in the Armed Forces, the second reaction can be the expectation that you are getting out of the military. Many people might consider the military a dangerous career choice, with its members facing harm and death on the front line. Although many service members must deploy and face the possibility of injury, and for some, even when they return home, they are dealing with combat trauma, that is not all we do. Most of us spend our time in uniform on military installations worldwide, just as far from battle as most Americans. While time away from family can seem long, especially when serving in a deployed location, it is just a part of what we do. I was stationed alongside thousands of men and women in uniform, and most of us spent only a part of our duty on deployments.

In my experience, very few people would ask a man if he were getting out of the military because he is expecting an addition to the family. But it is often the question asked of many servicewomen when announcing the arrival of a new baby. For some, it's more than a question it's a pressure to choose motherhood or service. For some, the answer is easy, choosing motherhood and separating from the military. For others, the answer is also easy, choosing both service and motherhood. Those choosing both are often excited, looking forward to the arrival of a new baby, and ready to take on motherhood and service to our country. It is a decision each woman must make for herself and her family. In the end, we hope to get the support we need from extended family and friends. But either way, we are ready to face the new challenge that service and motherhood would bring.

For moms in the military, we carefully weigh the decision to continue in the Armed Forces. We love our families back home. And while we hope to get support from our extended family, it does not affect our decision. Most of us are excited to share our good news with family and, mostly, they share in our joy, even if their happiness is not immediate or overshadowed by worry. But ultimately, we decide our career path, and our extended family's love and support are always welcome.

I can't really remember my family's reaction to my first pregnancy, most likely because I lived in Europe and communication was costly long-distance calls and "snail" mail. But I do remember feeling support. Their worries were more about me in Europe and so far away from home that if something happened, I would be alone. With the second and third pregnancies, my family realized I could handle it and didn't have any qualms. In fact, I remember my Uncle James told me not to get out but to keep going, which was a decision he wished he had made.

-Patricia Q., E-7, USAF (Ret)
First Sergeant and Mother of three

My family's reaction to my joining the Navy as a single parent was not as positive as I had hoped. Their preconceptions of women in the military were a bit negative. They gave me a lot of warnings, and they were not happy with my decision, but they felt it better than if I were going back to California. There were worried, and they felt the military was not a place for women, especially one with a child.

-Jacqueline G., E-3, USN Veteran
Personnel Specialist and Mother of two

I am not from a military family, so it was a big deal when I first joined the Air Force. My parents were like, "What?" But they could see my growth, even out of basic training. So, when my husband and I were expecting our first child, my parents were not really concerned because they knew I had a way to support my family.

-Anna P., E-5, USAF
Services Specialist and Mother of two

My family was glad we were in the military because there were medical issues, and we were in the best place for my son's medical treatment. My son had excellent care, which was definitely better for all of us.

-Carol G., USAF
Teacher and Mother of one

I knew my family would be happy with whatever I wanted to do, so I felt supported. I think any flack would have come from outside my family.

-Pagerine J., O-5, USAF (Ret)
Medical Service Corps and Mother of one

Honestly, I couldn't really tell you. I didn't really have that discussion with my family. It was like, hey, we are going to have a baby, and we're all excited. To frame this, I have no issues with going against the grain a little bit, not going with the status quo. Some people might be like, "Oh my God," whereas I am just like, "That's what I am going to do." They can get on board or not.

-Leslie L., O-5, USA
Human Resource Officer and Mother of twelve

In my immediate family, no one had any issues with me being a mom and remaining on active duty. There were other people...like one of my former pastors. He just said, "You need to get out of the military and stay home and take care of your children." He was an older gentleman, and in his mind, a woman's place was in the home.

-Yulanda B., O-4, USAF (Ret)
Personnel Officer and Mother of one

My mom asked, "Who will take care of the baby?" She could not believe I would leave them when required, but my husband was a phenomenal Mr. Mom. My children did not miss a beat. He stepped up and did what needed to be done, even when I deployed. When I asked if he would do this or that, he'd say, wouldn't you do it? And he did it and kept up our routine.

My brother didn't go so far as to call me a bad mom, but he did say that he couldn't believe I left my children to go to Iraq, where I could have been killed. I told him that I could have been killed crossing the street.

-Eboni B., E-8, USA (Ret)
Career Counselor and Mother of two

I didn't ask my family about it. I didn't come from that type of family. I had been married and gone for five years so I didn't ask my family and they didn't offer any opinion. My mother had lived with me while in law school and was close, but she didn't offer an opinion; they were unfamiliar with the military and had no idea what I was doing, so I didn't ask. On the other hand, my husband was a military brat, and his family saw nothing wrong with it. We were doing the same thing they did.

-Patricia H., O-5, USA (Ret)
Judge Advocate and Mother of three

My parents were actually very supportive. They knew I had a stable income and job. They were concerned about my education goals and wanted me to finish that before I had children, and I did. I graduated the same year my daughter was born. Of course, there were questions about what I would do in certain situations, but I had the support of my parents and siblings. I moved from Canada and joined the military as a Canadian. My extended family doesn't understand the military at all, so they were stunned when I decided to stay in after having my child.

My in-laws were not quite as supportive. My husband's family all live in the same place, and no one leaves, so we surprised them with our decision. They have always praised me for my service, but there have been comments about taking their son and grandchildren away from them.

-Toni D., E-8, USAF
Personnel Specialist and Mother of three

I think there was a lot of worry on their part. Mainly because my husband was away for training when our second child was born. Our first assignment was to Mississippi, and they were worried about me traveling with two small children and having to do it all on my own, and they wouldn't be nearby. And that is also how I felt at first.

-Ebony A., USAF Spouse
Mother of three

43

My mom wanted me home; a mother always wants her children close. So she wanted to be with me while going through my pregnancy. My sister, who knew what it was like to be in the military and be a mom, supported me in my decision.

It was kind of balanced. Some people said I couldn't do it, but others were supportive.

-Kimberly B., E-5, USA Veteran
Logistics Technician and Mother of four

My family supported my decision. They knew what was going on in my life and knew it was a way for me to provide for my daughter while my husband was not working. I also wanted her to have a strong female role model. On the other hand, one member of my husband's family said I was a bad mom for joining the military and "abandoning" my daughter. It was hard to hear, but I knew they didn't know what I was going through, so they couldn't understand my decisions.

-Jennifer H., E-3, USAF
Intelligence/Cyber Specialist and Mother of one

I had to fight my parents to join the military, but since I have been in, they have been really proud of me and supportive. And they were just ecstatic that I was about to have a kid.

-Chelsea D., E-7, USAF
Medical Lab Technician and Mother of three

Honestly, my mom, dad, my family...none of them actually wanted me to join. They would make comments like, "You're leaving your kids, so you are saying your kids are not important." They were not supportive at all. My dad has since changed his tune. They would also ask, 'If you go to war, who will keep the kids?"

-Mykeia L., E-3, USN
Logistics Specialist and Mother of five

I really shocked my family with my decision to join the military. And my husband and I decided not to say anything until we knew I was going to join; I was already a wife and mother. Once I told them, my dad didn't say anything, but my mom quickly flew off the handle. My dad didn't understand my decision either. My husband's family was way more laid back and accepting of my decision.

My kids were five and eight, so my parents were attached to them and just weren't sure. Even now, my mom is still not too keen on my decision.
-Kelsey Q., O-2, USAF
Operating Room Nurse and Mother of two

My family was very supportive because I was older, and they knew that if it was the right decision for my husband and me, we would make it work. I think they realized that we were on a mission and would do everything possible to make this work. My husband was already retired and gave our family a sense of stability.
-Cynthia S., O-4, USA (Ret)
Nurse and Mother of two

I have to be honest; my parents thought I was crazy to adopt. They thought I had a great life and didn't understand why I wanted to adopt a child. So once I got the impression that they thought what I was doing was crazy, I really didn't mention it again until she was home and legally mine. And of course, then they met her and, who can't love a baby?

I think that one of the reasons I waited so long was because I was always worried about what my parents thought. That they thought I would have to be married and whatever, but at some point, I thought this was ridiculous, and I needed to live my life before it was over.
-Melissa B., O-8, US Coast Guard
Chief Counsel and Judge Advocate General and Mother of one

They were supportive, but there were a few things we disagreed on. For example, my mom stated that no grandchild of hers would be in military childcare. So I had to explain that my husband and I had to make the best decision for our family. But I felt very supported.

Regardless of who you are listening to, it really is up to the member (and spouse) to make the best decision for their family. Everybody's decision is for different reasons, and for those who choose to separate after pregnancy, I do not think less of their service because they had to make the best decision for them and the best way to move forward.

-Janell H., O-3, USMC Reserves
Logistics Officer and Mother of one

I am my parents' first born and am very independent and driven. Once I set my mind to something, I just do it. I have been fortunate that when I decided to do something, I had my family's support to reinforce my belief that I could do it. But, I have some grandparents who were unhappy with my decision to join the Marine Corps. One was like, why not join the Air Force?

And when I got pregnant, I did have my family's support for my decision to continue in the Marine Corps.

-Corey G., O-3, USMC Reserves
Combat Engineer Officer and Mother of two

For us, it was a switch from the traditional roles. I am in the Corps, and my husband gave up his challenging and promising Marine Officer career to become a stay-at-home dad for now. Even our church family was surprised by our roles in our family. My husband loves having the opportunity to spend so much time with our boys. For him, some days are excellent, and spending so much time with the boys is great. At other times...as the man and the head of the household, he had to make some adjustments.

-Maria F., O-3, USMC
Training Officer and Mother of two

I didn't really have any expectations for how my family would react; my family is not a military family. I think one of my grandfathers served in the military. So they didn't have any expectations of me either. I was just going to stay in; it was not a question for them or me. So I really felt like I had their support. Even my in-laws, whose parents were in the military, were so happy. I was the first to have children, so ours was the first grandchild on both sides, and they were ecstatic.

-Jennifer P., E-8, USMC
Administrative Specialist and Mother of two

I don't think anyone was surprised, and I don't think there were expectations that I would get out. My mom was a working mom who worked her whole career. Not in the military, but she worked her entire career. I don't think anyone thought I was getting out because I was pregnant.

-Rojan R., O-6, USAF
Program Manager and Mother of three

CHAPTER 5:
IMPACT OF RAISING A CHILD WITHOUT EXTENDED FAMILY (ON MOM AND CHILD)

It is not so much our friends' help that helps us, as the confidence of their help.[13]

-Epicurus

Even today, many of us depend on our extended family for support. Support from family, family friends, and our lifelong friends: our "village," if you will. I know people who have spent most of their lives in the same area where they were born and/or grew up. When I traveled back home, I often had family and friends still living in the same neighborhood and was not interested or concerned with moving away. My cousin thought my lifestyle in the military was crazy and not something he would want to do. While modern technology has made contact across the miles easier and simpler for those living so far away, there is nothing like seeing and being with one's family.

Unfortunately, life in the military sometimes makes access to this immediate support difficult, if not impossible. Being in the military puts you in the same situation as your fellow Airmen, Soldiers, Sailors, Marines, and Coast Guardsmen. And no one can really understand your situation like someone in your situation. Some of us often redefine their definition of "family," which has nothing to do with your DNA and more with the people around you. We can and do create our "family" wherever we are to feel the connection and support we all need at times.

It is not just major events, like childbirth, which can make the distance so obvious. It's missing birthdays and holidays, family get-togethers, and other milestones, and even having to mourn by yourself when someone back home dies. Most of the time, you can go home to attend the funeral, but reality and sadness can kick in when you are back to day-to-day life after the funeral. Worse yet are the times when duty does not permit you to make it home in time to say goodbye or attend the funeral. This is often when your fellow service members step in to fill the gap. To help you shoulder the grief and share the pain of separation.

I remember when my great-grandmother died, and because of duty and location, I could not get home to attend the funeral. My friends helped me during this time when I felt so alone and far from home. I often found solace in my fellow Airmen; we developed bonds that were, in many ways, as strong as the bonds of family.

I found that same to be true for my fellow service members. For most of us, the camaraderie was strong and comforting; even today can bring a smile to our faces. Knowing you are not alone, that someone can understand and will support you is priceless. Who else can understand the life of service members? Who else could empathize with all the ups-and-downs you go through in the Armed Forces? Your fellow service members. Developing these friendships eased family separation and, for some, became the family we needed or the family we never had. Through the years, many friends laughed with us, cried with us, celebrated with us, and commiserated with us.

While nothing fits or feels precisely like the family you grew up with, creating your own "village" can fill the void and become as close as family.

One of the consequences of service is that your children may not grow up with the same family dynamics as many of us. No grandparents down the street or across town or cousins to spend the night with. Unfortunately, this can also mean birthdays and holidays without the extended family and no day-to-day interaction with that same family.

Many of us create our family right where we are. A surrogate family to fill in the gaps, not just for yourself but for your children. It could be co-workers, church members, neighbors, or even the tightness of your family nucleus that you value and learn to nurture. It's different for each of us, but what is the same is that we strive to create a family where we are, helping us and our children grow and develop. This often helps ease the task of raising a child without the immediate access or assistance of the extended family... for both the servicewoman and her child(ren).

Being a young Airman living in Spain when I had my first child was challenging without my family. I know that most of my siblings lived relatively close to our parents and had that support. So, as a new mom, it was a struggle. Especially living in Europe before email, FaceTime, etc. Calls were expensive, and I had to find a new support system. Because I was raised to believe that it takes a village to raise a child (which was certainly my experience growing up), I found mine where I was. It included co-workers and friends. I remember worrying about my child in a group daycare setting, and a friend's wife (a military spouse) agreed to help me out and ended up providing daycare for my daughter for almost three years.

And with my second and third, it was the same, I had to find my "family" where we were stationed. I continued to depend on

51

my military family. Again, living in Japan, I had to have someone who would be responsible for my children in case of evacuation of non-military personnel, and again, it was a co-worker's spouse who agreed to help. Fortunately, it didn't come to that, but I had my support system, and it gave me peace of mind.

I am sure that my service had an effect on my kids. Not growing up around family was, in my mind, a sacrifice they made. And there were times when I knew they felt like outsiders around their own family. For me, that brought a certain amount of guilt. But looking back, I think my children are stronger because of our military life, and I can see that they have forged their own relationships with the family.

-Patricia Q., E-7, USAF (Ret)
First Sergeant and Mother of three

My family was still a part of my life (after joining the Navy); we have a strong family unit. My mom had my son while I was in training and provided a lot of support. Even with the distance, I felt their support, and I don't know if I could have done it without the help of my family. My mom happened to be in Mississippi while I was stationed there and was able to assist. I really could not have done this without them.

Being in the military really helped me to grow up and look to myself for what I needed. I had to transition from child to adult quicker than most of my counterparts back home.

-Jacqueline G., E-3, USN Veteran
Personnel Specialist and Mother of two

I don't think it really impacted me. I joined the military at such a young age, and it really forced me to grow up, and in doing so, it made me more independent and not really need that immediate support around me. But, I feel like the places I have been stationed and the distance from home have kept me from visiting family as much as we would have liked.

My daughters have never lived around family, and the military has been their only lifestyle. I think they have grown to be resilient and learned to develop friendships to take the place of my family members. I noticed that sometimes when we go home, my oldest prefers to be with her younger cousins because the older ones (closer to her age) have their own friends and things to do.

-Anna P., E-5, USAF
Services Specialist and Mother of two

It was hard to be away from my family. My son had medical issues, and it was hard not having them to help out. I am sure we would have had much more support at home, but the best medical treatment was here in Texas. My family was not flyers, but they did come to visit. The separation from family was the worst.

For me personally, I really missed my family and didn't like having to travel so far to be with family. I didn't get to see my nieces and nephews grow up. And even after retirement, we chose to stay here (away from family) to ensure my son got the care he needed.

For my son, it was hard that he didn't get to grow up with his cousins. We went home often, but it wasn't the same. And because of his needs, we didn't move again. In that sense, we were unique. I can't say I missed out on the moving experience because I was so consumed with taking care of my son.

-Carol G., USAF Spouse
Teacher and Mother of one

Now, this was tough, especially my divorce. It was pretty challenging. I had just gotten my commission, and it was just my daughter and me. Fortunately, my first assignment as a single parent and Medical Service Corps was at the exact location where I had some family stationed. Sometimes you can feel like you have to do it alone with no one to depend on. You learn how to rely on yourself and just suck it up. You miss out on the family celebrations and holidays.

And I think part of it is that you grow up developing bonds with family. And then, when in the military, you go home and find you don't have those same bonds with the newer generation. So you can feel like an outsider with them.

For my daughter, I think it was definitely felt. She is an only child, and she probably felt alone as well. She wasn't overly excited to see her cousins at family gatherings because she didn't grow up with them and build those bonds growing up. I think she missed out on that part. It was beneficial for my daughter when we were stationed at the same base as my sister; for those cousins, she could develop that family bond.

On the other hand, she developed a tremendous sense of independence.
-Pagerine J., O-5, USAF (Ret)
Medical Service Corps and Mother of one

For me personally, I think it has to do with my not being prideful. If I need help, I am going to ask for it. I am big on developing relationships. For instance, my church, my sorority, and other organizations, and I have always been so fortunate to have a fantastic support network. I have been fortunate enough that others in my life say, "What can we do?" or "What do you need?" Most times, I was good to go, and when I needed something, I just happened to have just what I needed. I am a woman of faith and believe that God will always provide. I have lived in a homeless shelter, wondered where my food would come from, and shared a twin bed with my mom for a long time. I am not a stranger to struggle. And so hence, I am not a stranger to asking for help.

Because my mom was a single parent, I had to be independent, and she empowered me and ensured I knew my strengths and potential. And she really prepped my path to self-sustain. I always knew I was going to go off into the world and do something. So as long as I was in touch with my family, I was good. And I have a support network that I call my family, but they do not share my DNA. That is how I chose to look at it. My family is those who have been in the trenches with me, seen me cry, seen my family's ups and downs, and been there to support me. That is family to me. So, not having my immediate family was not a big issue. We love our family, and we stay in touch, and technology helps us even more.

For my children? Now, I will say that one disadvantage for my children has been that they have not had the opportunity to be with their cousins and forge those relationships. We have not been stationed close to family, but we are good. My kids are not missing a beat. And I think how you are as a parent and what you display will jump off to your kids. If you don't worry, they won't worry. When we do get together, we all catch up. For my younger children, now that we live near family, they get more of that. But overall, they are good.

-Leslie L., O-5, USA
Human Resource Officer and Mother of twelve

I was very fortunate because my mom lived with me for the first couple of years after my son was born, so I had that hands-on support from her. But I felt bad because my son didn't have the chance to be with his extended family. So, his extended family became friends who were not blood relatives. I had to be intentional when it came to him interacting, and we made it a point to try and go home at least once a year. It was a challenge. For me, I grew up around my cousins, and my son didn't have that opportunity. But I feel like technology has made it easier for us to communicate and stay in touch.

-Yulanda B., O-4, USAF (Ret)
Personnel Officer and Mother of one

I think it was more challenging in the beginning because, as a new mom, there is so much that you don't know. And sometimes, in the beginning, you need someone to "see" you when you don't know what you need. You are home and trying to do so much before you return to work, and you wear yourself out. Now, had I been home with family, I would have had someone to step in and say, go sit down and take a break. I didn't have that. I cried many nights because I felt I couldn't do it all and was overwhelmed. That was the most significant effect.

On the other hand, I always had my church family wherever we were. Even to this day, we still keep in touch with those friends. I have a friend, and because of our relationship, I have been in her daughter's life for eighteen years. Our friends become our "village" and take up the slack. In the military family, we have to be there to support one another because we are not near family. It's just a matter of finding someone you are comfortable with.

For the boys, there has definitely been an impact; we tried to mitigate it when we could, but it affected them. We spent the first eight years of their lives overseas, and we came home once during

55

that time. And it was in an age where you could not FaceTime. So, my boys were coming back culturally different from their extended family. I remember one instance when my kids were going on about the World Cup, and my cousins' kids were like, "What?" Or not knowing about a well-known restaurant because it wasn't in Germany. Knowing that my boys are not as close to my family was a struggle.

-Eboni B., E-8, USA (Ret)
Career Counselor and Mother of two

When I was a student/spouse, we were close enough to go home quite often but moving to Germany was a whole different thing. I didn't even really know how to cook when I got to my first assignment in 1992, and, at that time, I couldn't call home. It was learning to do things and balancing my life. My husband was further along with his career, and I had to figure out some things. I grew up during that first assignment. I thought I was grown before, but it was then that I grew up.

But I was able to find friends, and we became family. One of my co-workers became close, and even our children became friends. So you find your "family" and I did; we did things together, even traveled together. That is the beauty of the military; you find friends that become family and develop lifelong friendships.

In the beginning, my daughter was so young that I don't think it was an issue for her then; she didn't know the difference, and she enjoyed the "family" we adopted. We all did so many things together. My other two were born after I joined and didn't know life outside the military. As they got older, they didn't like the fact that we often moved and did not see extended family more regularly.

-Patricia H., O-5, USA (Ret)
Judge Advocate and Mother of three

It was definitely challenging, especially with my first child. My husband was a teacher and coach, and his hours didn't allow for him to help out in the mornings or afternoons. In addition, I was assigned to an ROTC detachment; I did not have the convenience of living on or near a base. As a result, finding daycare at 6:00am was hard; most places were not open. I was fortunate enough to be able to take the baby with me to PT, and then I could take her to daycare.

I am so thankful they allowed me to do this because I didn't know or trust anyone, and there was no family close enough to help. My husband's family was about an hour away, which was helpful for weekend visits. Unfortunately, since we have moved, we don't have any family nearby, which has been an adjustment.

I loved having and knowing my family, but it's not there for my kids. I think the hardest thing is that they do not have the same type of relationships I have with family. My family travels, and coming to visit is easy for them. My husband's family does not travel much, so they don't get to see us as much. And with the pandemic, we haven't been able to travel. And my youngest has family she has not yet met, and it's hard knowing that there is a whole side of her family that she hasn't met yet.

For that first move with my daughter, it was difficult for her. She was two and a half and was used to seeing her grandparents, and we moved to a place where we didn't know anyone.

My children are young (six, three, and one) and don't really know anything different than how we live. When family comes to visit, there is all this hype about who's coming, and then they get there, and the kids don't know them. It takes a couple of days for the kids to adjust, and then they're leaving, and the kids are sad. I think that has been the most challenging thing...not having those relationships.

-Toni D., E-8, USAF
Personnel Specialist and Mother of three

It was really tough, actually. After my second child and our first assignment, I had post-partum depression, we were away from family, and my husband worked nights. We had a toddler and an infant, and we were in a place where I didn't know anyone; I wasn't working, and I spent a lot of time at the mall just to get out of the house, so he could get some rest. Then I finally met some of the other wives and learned about spouses' resources, which helped.

For my kids, I know it was very different from what I had. No one in my family moved; I went to the same school my mom and grandmother attended. We lived in a small town and had our whole family there. So I know they missed out on having that relationship with the family. But I think it made them more resilient and helped them transition from high school to college, for example. I believe they are more independent and have skills for making friends and persevering. This has made them close to one another; they knew they had each other, even if they didn't have anyone else.

-Ebony A., USAF Spouse
Mother of three

Before joining the Army, I lived with my older sister, who had already joined the Marine Corps. I lived with her to help raise her son, and leaving them was especially hard because I developed a strong bond with my nephew.

Being in Germany with my first child, I found a group of friends that offered that support, and they became like my family...my military family.

After I divorced my husband, I had the added stress of being a soldier and a single parent. And even though we were not living in the same place, my sister was a source of emotional and financial support for me, especially after my divorce.

Being in the military forced me to grow up; I had more responsibilities and became more disciplined. In addition, I had to take care of my family, and being away from my extended family gave me separation anxiety. I missed being with my dad, with whom I had always been close.

The overall impact of serving has strengthened me in some ways. It helped me to know what to do and how to respond to the unexpected without depending on others. However, it also caused me some issues; I suffered from anxiety and depression. At times, I felt alone, like no one understood what I was going through or what it took to be a full-time soldier, a full-time mother, and still take care of everything. In addition, a financial struggle resulted in me working two part-time jobs to make ends meet. This also influenced the time I could spend with my children.

My struggles were also felt by my daughters. They didn't get to know their extended family while I was on active duty.

-Kimberly B., E-5, USA Veteran
Logistics Technician and Mother of four

It's been, at times, a struggle and frustrating, but it is doable. You just learn to find a community of support; for me, it's been my church. They have become my second family. The people I work with have also been supportive and understand my role as a single mom. I have been fortunate in that way. My daughter is really young, and I worry about her being away from my family. But she has been a trooper, and this has made her stronger. She has already been meeting new people and experiencing new things. So far, this has had a positive impact on her.

-Jennifer H., E-3, USAF
Intelligence/Cyber Specialist and Mother of one

I believe it takes a village to raise a child, and I have been extremely lucky to network and find individuals to help me along the way. Even with my family care plan, I have a civilian co-worker on my family-care plan. I am not afraid to ask for help when I need it. I struggle at times, but I have been very fortunate.

Other than that, it has been great. The military has allowed me to grow, travel, and branch out on my own.

For my children? They are so young. But it's been hard on my parents since my children are the only grandchildren. But for this assignment, we only live a few hours away, so we have been able to get together, COVID protocols permitting.

-Chelsea D., E-7, USAF
Medical Lab Technician and Mother of three

It was really difficult to be away from family. I now have five kids and one with Sickle-Cell, which can be hard to manage. When he gets sick, it can take weeks to recover, and not having my family around can be really hard on me.

For my kids, the separation from extended family, can, from time

to time, can be challenging. My oldest remembers living near family, and they do miss their grandparents and cousins and want to go and visit, and I can't say, let's just go visit your grandparents.
-Mykeia L., E-3, USN
Logistics Specialist and Mother of five

We were so fortunate back home that my mom and mother-in-law were around to help with daycare. After joining, we had to make an adjustment. Cost of daycare for one and for the kids not seeing their grandparents every day for another. It was an adjustment for my daughter, definitely. Over the summer, my husband stayed home, which was helpful; once she started school, she was much better. I know I couldn't do this without my husband; he's been great.
-Kelsey Q., O-2, USAF
Operating Room Nurse and Mother of two

We were fortunate that wherever we were stationed, we connected with people who became our extended family. I think that is the joy of the military. You just run into people who were Godsent for you. We are all in the same boat and understand what we were going through. When our children were younger, I missed having family near us.

Our family was always important to my husband and me; for our boys, we made it a point to travel home to see family once or twice a year so our kids could get to know their families. We always tried to keep that connection for them. We also made it a point for them to stay connected to the family through visits and phone calls. But honestly, they were so young that I don't think they remember as much as I do.
-Cynthia S., O-4, USA (Ret)
Nurse and Mother of two

It's kind of crazy, but when I first adopted and later got married, my husband was still living in Florida, and I was transferred to Colorado Springs. It was horrible for him to be away from us. Whenever I am away from our daughter, it is hard; I could only imagine how it was for him to live so far away.

We are not the typical military family in that we have access to financial resources that some military families may not have access to. Our resources allowed me to have someone living with me, helping with the baby for the first six months or so, and I still do now and again. That was a huge help. The Coast Guard is different from other services; we are dispersed all over the place in small groups. They don't usually have housing or daycare or that sort of thing. My only joint tour was at Peterson AFB, and there was a great daycare center within a mile of my base housing and work. It was great and made my life much easier but not typical for Coast Guard.

My daughter is still young, only five.
-Melissa B., O-8, US Coast Guard
Chief Counsel and Judge Advocate General and mom of one

When I had my first child, I was stationed on the east coast, and my family (mine and my in-laws) were on the west coast. So my parents and in-laws took turns coming to visit for the first six months after I returned to work. That was great because my husband deployed shortly after my son's birth, and it was nice to have that extra support in the beginning and not have to do everything by myself. I also had four older siblings with kids to look up to and call with my questions.

Being away from my family, I had to find ways to be creative in caring for and supporting myself. I used a lot of sites for help and used my friends. I remember one morning when I needed to report for a unit hike at 4:00am; I remember having to arrange for someone to meet me in the parking lot to do a car seat swap so that I could make it to the hike on time. Others don't always understand what mothers must endure just getting there. I also had to band with other military moms and set up that support network. I sometimes think that just being able to see other women achieving the same thing encourages me to keep moving forward.

Also, my support network (tribe) included military spouses who were also moms. Sometimes, there is the perception of a rift between servicewomen and spouses who are moms, but when it comes down to it, we really band together and support each other.

For my children, we were able to spend holidays with family. Now that we are in California, we can see family more. Not being around extended family has made our family traditions and activities more important and powerful. The structure you provide creates an anchor for your family when you can't be with extended family.

-Janell H., O-3, USMC Reserves
Logistics Officer and Mother of one

Becoming a mom is a whole mind shift. Typically, women in the military are pretty independent and confident, and if you are me, you hate asking for help. It was really, really hard not having that support. My first birth was traumatic, and fortunately, my mom was able to come up for a few weeks to help. But after she left, I still had the baby and a new life. At that time, my husband was working pretty hard, and I was trying to adjust to a baby that didn't sleep well. It really wore me down. It is so hard not being located near family to help out. We were lucky enough to find another dual-military couple that we could lean on. We became that support for each other. And I know we will be close to this family forever.

It did impact my children, but it's the only lifestyle they know. We tried to visit as much as possible. But now that we are closer to family, there is a vast difference. It is so different; my kids know the family and immediately created this bond. I don't know how it happens, but it does. And it is so healthy for my kids to have this relationship with the extended family.

-Corey G., O-3, USMC Reserves
Combat Engineer Officer and Mother of two

It's been tough. Fortunately, my mom came out to help us with our second child. And my father-in-law is retired, so he has also been able to come out from time to time. But, it has been challenging trying to find that support, especially when I have to be away from home for work. Trying to find that support is probably one of the hardest things. Our friends back home have that extended family in the same area to help out and are always asking us how we do it. We don't know how but we just do it.

Because my children are so young, I don't think they know any difference.

-Maria F., O-3, USMC
Training Officer and Mother of two

So, it was hard. My first child was born in Virginia, so it was a driving distance; not close but not too far, and that was good. But my parents and in-laws were working full time, so it wasn't very often that they came. With my second son, we were in California and closer to some family, but not much.

We are a dual military family and didn't get babysitters, have date nights, or go on vacations. We were very schedule oriented and just got into a routine. But, of course, the mom is expected to take care of everything with the kids, like medical appointments, school events, sports, etc. My husband had no idea of all the little things needed to take care of the kids, and he expected that I would take care of these things. But there was also an expectation from my command that my spouse would be doing some of these things.

I could not do all the crafty things that other parents might do, drop them off and pick them up right before and after school (they are in before- and after-school care), and be active in PTA and still be successful professionally.

We are in our third move since having kids, and I am finally at a place where I have that support system with our neighbors, who are primarily military families; they are amazing, and I trust them with my children.

But that's the life my kids know. They don't know another lifestyle.
                                        -Jennifer P., E-8, USMC
                            Administrative Specialist and Mother of two

You know, I didn't think much of it. I don't know what it would have been like to have my parents close; I only know what I had to do. I figured it out. I don't think there was ever an expectation that we would always live close to one another and be in everyone's circle all the time. It wasn't really a big deal. For holidays and even birthdays, we figured it out. I used the CDC for my first child. It was excellent childcare; it was close to work, and I could see him at lunch and do all the activities with him. I also found a teenager to babysit, and I didn't think much of it. When I was a teen, I babysat; you know, it was kind of like the "circle of life." I didn't find it weird or unusual or think it was different.

63

It was a little strange or different when I looked back at all my high school friends who lived in the same neighborhood and on the same street with the same stuff. I was like, "You guys never left town." They were probably looking at me like I was out there all by myself. So I think it is all a matter of perspective and maybe your sense of adventure or what you are willing to do. I never really look at anything as a negative.

I am huge on "it takes a village," and I talk about that and encourage people. I call it your community; you have to make your community wherever you are. What really irks me when moving with young children is the schools always require two or three emergency contacts, and you don't know who to put down. It's such a source of stress that it can make you want to cry because you are by yourself or just you and your spouse, and you think, who am I going to put down as contacts? If you are not an extrovert or someone who puts their hand out to meet new people, it can be hard to figure out who to put on that sheet. It's one of those situations where a mom can feel overwhelmed when you just moved, and they expect you to know people. But it's a time to say okay, maybe I have a female boss or a boss whose wife can help out. Take a deep breath and know you can get through this; this will not be the thing that pushes me over the edge. That's always the goal: to not let that one thing be the thing that pushes you over the edge.

For my kids, I don't know. It's like one of those questions, "If you had taken a left instead of a right turn, what would your life be like?" Our families have been involved in their lives as best they can; my kids know their families and have a life with them. Would it have been different if we lived next door? Sure, but not being next door to each other, I don't think they felt less love or had less access to things. They also got to see a lot more states and countries. It's pluses and minuses in their experiences. I think they will be productive citizens, which is my goal. Hopefully, by living in different places, they have experienced different cultures, people, and things of America, and hopefully, they will turn out okay.

-Rojan R., O-6, USAF
Program Manager and Mother of three

CHAPTER 6:
THE IMPACT OF EXTENDED ABSENCES (ON MOM AND CHILD)

When you are a mother, you are never really alone in your thoughts. A mother always has to think twice, once for herself and once for her child.[14]

- Sophia Loren

Today, service in the Armed Forces almost guarantees that there will, at some point, be a separation from your spouse and children. Even more so now than in the past. The separation can come in the form of deployments (which last up to eighteen months), remote assignments where family members are not authorized (one to two years), or temporary duty assignments (TDYs), all of which result in separation from your children (for days or even months).

All service members experience one or more of these types of separation. We all know that it is "a part of the package" we signed up for, and we do it with willingness (most of the time) and a sense of pride. However, because women tend to be the primary caregiver for their children, these extended absences can be more challenging for servicewomen with children. It can mean rearranging schedules for your children, temporarily relocating them, missing birthdays, the first day of school, recitals, games, and everyday happenings for our children. Like having to go TDY right after getting to a new assignment before your family even has a house and is still living in Temporary Quarters. It happened to me; my husband and children had to get settled into our new home while I was in training. Was it the home I would have picked? No, but it was home, and in the end, it was a good fit for us.

Sometimes, the separation might be a welcome change. It can also be an opportunity to allow your children to bond with dad or other family members, or even a time for us, as moms, to recharge and rejuvenate. A time to focus on ourselves, which ultimately makes us better moms and servicewomen.

Like any mother facing separation from her children, we try our best to prepare our children for the break and work hard to ensure that we are there for them when we are home. We strive to let them know how much we love them and reassure them that we will be back. More times than not, through these absences, our children grow and find the strength to deal with our absences, sometimes even better than we do. Sometimes, they learn just how independent they can be; they learn how to care for one another or relate better to the caregivers who fill in our absences. We are often surprised by their resilience.

I think many of us worry more about our children than ourselves when facing such separations. Even though we, as adults, can see the big picture and realize how short the time away is in our overall lives, we do worry about how our absences will affect our children. Because we can deal with these separations in ways that our children may not be capable of doing just yet, we worry. But we do our best to help them understand the importance of what we are doing and try to be as accessible as possible, which has been made easier through social media, email, texts, and video calls. More importantly, we want them to know we love them and that they can and will be okay until we return.

Going away is not always easy, but when you keep the mission in mind, know that your children will be all right, and focus on the growth that everyone could experience, the TDY, deployment, or the remote assignment is easier to deal with and get through.

I did not have the opportunity to deploy but went on many TDYs ranging from a few days to a few months at a time. I learned to take it in stride, as part of military life. I knew my children were well cared for, and I could go and come back without too much heartache. Sometimes, it was a welcome break from the day-to-day operations of work and home life. Of course, some were harder than others, depending on how old my children were. I remember leaving when my youngest daughter was not even one, and I had to leave for just seven weeks. It was rough. Maybe because she was so young, my motherly instincts and post-natal body and mind were screaming to be with her, or maybe the fear that she would forget me. Which she did, or at least that's how I felt when I got back home, and it was days before she would let me hold her. But I also remember leaving my children in Japan while traveling back to the states. I can honestly say that I felt confident in their capabilities and could do what I needed to, knowing they were doing well.

Each time I went TDY, I knew that my children missed me, but I am certain it gave them the chance to bond with their dad and learn just how strong they were. Looking back, I know the impact was much more positive than negative.

-Patricia Q., E-7, USAF (Ret)
First Sergeant and Mother of three

I did not have to be away for temporary duty while on active duty and didn't have to spend time away from my son other than my initial training and a few short trips. He was so young that I don't think it impacted him very much, if at all.

-Jacqueline G., E-3, USN Veteran
Personnel Specialist and Mother of two

Deployments and time away from home are hard, whether you are military or civilian. Still, my first deployment after having my first child was the hardest. Trying to separate and remind myself why I chose to stay in the military. I never wanted to leave them, but it always came back to, "This is what I chose, I am glad I am doing it, and it's bigger than just me."

My kids are so resilient. I am sure my time away has made them more independent, especially my oldest. She had to grow up quick. I can see a change in them now. Each time I leave and return, I see their growth and development, even when I could not be there.

-Anna P., E-5, USAF
Services Specialist and Mother of two

My husband had to do a remote overseas assignment for one year when our son was in the first grade. It was actually okay. I worked at the school my son attended, and my co-workers were a super support system, which helped me get through. However, because of my son's medical issues, my husband's absence could be nerve-racking, and I prayed there would be no serious issues that I had to deal with on my own because my husband was not here in the states with us.

My son is on the autism spectrum, so I don't think he really realized that his dad was gone for that long. And I guess that made it easier for me too. I am sure it was harder for my husband to be away from us for that year.

-Carol G., USAF Spouse
Teacher and Mother of one

The guilt, the guilt, and the guilt. I remember one deployment, my daughter's grades dropped, and she would tell me she was worried about me. The responsibility for making her feel guilty was challenging. As an only child of a single parent, it was always just us, and when I was gone, it was hard.

I remember missing a birthday, and that was hard. And I think, sometimes, we overcompensate for those absences.

For my daughter, I think it was tough. Since it was only us, I think she may have felt a bit lost when I was gone. And she would tell me that she was worried about me deploying, and her worrying was stressful for both of us.
-Pagerine J., O-5, USAF (Ret)
Medical Service Corps and Mother of one

There is the political answer, and there is the truth. My truth: Let me tell you something, I deployed for six months, and I enjoyed myself, minus the things that come with deployments. And you know why? Because I had been wife'ing and mommy'ing for so long, at that point, for like fifteen years, I deserved to have some Leslie time. I have no guilt or shame. I knew their dad had it, a support system was available, my children knew how to respond and react, and we had the technology. It was all good; the separation did not affect me. I enjoyed my time away and could get in shape; it was great. I missed them, of course, but I was okay.

For my kids, from what I can tell and from conversations we've had after my return, I don't think there were any adverse effects. I guess because of my attitude and the ability to communicate with them. There was healthy communication, and I think they understood that it was okay for me and that I was serving my country. And, because I have such a robust support system, and they stayed so busy, they didn't have time to worry about me being gone. And I was careful not to put the weight of what I was doing on them.
-Leslie L,. O-5, USA
Human Resource Officer and Mother of twelve

So, for me, my mom was with me except for when we were stationed in Germany. I deployed twice while in Germany and relied heavily on my friends (my extended family). My husband doesn't cook, so I was a little concerned. But I had made arrangements with my close friends who could step in and help. In fact, I kept a friend's child while she was deployed. You use the resources you have. I tried my best to focus on the mission and be intentional. I devoted time to Skype with my husband and son so I could continue to be a part of their life. The worst thing that happened was the first time we Skyped. My son was crying to hold my hand; it was emotional. And I tried to ensure that I could at least be home on his birthday when I had to be away. And with four deployments, I was always home for his birthday. Not all holidays but for every birthday. I also made it a point to try and stay involved. I would even email his teachers during my absences to stay abreast of what was happening and stay as involved as possible as if I were still home.

For my son, I think it made him even more resilient. I remember my last deployment to Afghanistan, and I remember talking to him about where I was going and the danger of the situation. Then, of course, he was older and said, "Mom, you got this."

-Yulanda B., O-4, USAF (Ret)
Personnel Officer and Mother of one

It can be hard when you leave and come back and try and step back into their world. My husband was always hands-on with the boys, so when I had to be away, he was still there to care for the boys. But I remember after my first deployment, coming home and feeling out of place. He had everything running so well, and I wondered, "Where do I fit in?" I wanted to come back and be "mom," but their life was going on with their routine. We had to sit down and talk to my husband, and I was like, hey, let me in. I remember us talking about reestablishing our roles. I also feel like because I had my children later in my career, I had a bit more freedom from work to be there for them.

Because my husband was so good and active with the boys, he did everything he could to keep their routine going so that they were not affected by my absence.

-Eboni B., E-8, USA (Ret)
Career Counselor and Mother of two

We were a dual military family, so we both deployed, and the two times I deployed, he also deployed. Today, my children don't feel that the military is for them after seeing what we have had to do and experience.

When I went on short-term TDYs, it was hard on my son. He seemed to get in trouble every time I left. For one deployment, it was hard because my daughter (a senior) demanded that I still be as active in her life. Of course, by then, we had email, and she was constantly in touch and wanting things.

During one deployment, I was seriously in harm's way, and my security team protected me because I lost focus. I didn't tell my children until much later, and even years later, they were glad that I told them before I shared it with my congregation so they would have time to process it and not in front of everyone.
                                        -Patricia H., O-5, USA, (Ret)
                              Judge Advocate and Mother of three

When I am away, I just miss them. It breaks my heart to leave my children. I have been on one- or two-week TDYs but have not had to deploy since having kids. It's tough when I am gone if one of them gets sick while I am away. I just feel bad not being there to take care of them. Thankfully, we have the technology to keep in touch and see them often while I am away.

I think it has impacted me more than my children. Maybe because they are so young. What is helpful is that my mom has come out for my absences, and they have really enjoyed spending time with their grandmother.
                                              -Toni D., E-8, USAF
                              Personnel Specialist and Mother of three

Those times when my husband was gone could be stressful, depending on whether I was working full-time or part-time. The kids were very active with scouts and sports, and I was always running around trying to keep their schedules the same. It seemed like I was always coming or going: going to work, coming home from work, going to pick up the kids, or dropping the kids off. And I am terrible about asking for help because I wanted to do it all even though I had a support group. I felt this lifestyle was my choice and had to do it myself. But there were times when I had to break down and ask for help when my husband was gone. It was stressful, but it made me appreciate him more when he was at home. I could definitely feel the difference when he was gone.

I know it was tough on the kids, and I tried hard to keep them busy. I was really thankful for the technology that helped us stay connected. He would Skype with them, so I think that made it a little easier. He also brought them something back, giving them something to look forward to. Our youngest doesn't remember her dad being deployed. She was two, and in the beginning, it was really hard on her, and then by the time he came back, she didn't remember him. So that was an interesting experience.

I think overall, it has been positive. It has made our children resilient, and for my daughters, it has helped them see what can be done when needed; I think for my son, it has helped him better understand what a mom goes through. But, at the same time, I think it has made them worry more about me because I have a medical condition that sometimes prevented me from doing things, and we had to have help.
-Ebony A., USAF Spouse
Mother of three

I remember one deployment when my daughter was two years old. At that time, I was taking care of my nephew while my sister was deployed, and I got orders to deploy to Cuba for several months. I didn't feel like I could take the children home, so my mom came out to Colorado to help. When I returned, my daughter was like, "Who are you?" I felt like my daughter didn't even know me. That impacted me in a significant way. A few months after my return, I was sent to Korea on a remote fifteen-month assignment. I missed out on a lot of her firsts. I just broke down; that was very hard for me; at that time, depression and separation anxiety set in for both of us. Detachment issues for my daughter caused some real problems for me.

After spending fifteen months in Korea, I separated from the Army a few months into my next assignment in Georgia. For my sake and the sake of my children. I knew I was not willing to leave my child again.
-Kimberly B., E-5, USA Veteran
Logistics Technician and Mother of four

I haven't had to be away from my daughter except during my training, which was about seven months. Leaving for basic was pretty hard; probably one of the hardest things I had to do. But I knew it was for a short time and that, looking at the big picture, I could get through it. My husband sent me photos at the time, which was good for me. At tech school, my training had to come first, and I had to be a trainee first and a mom second. Just knowing that it was temporary helped me get through. Luckily, my daughter was in the area, so I could see her during my training. Due to the COVID pandemic, traveling has been reduced, so I haven't had to go on any TDYs. During and since my training, my daughter seems to have adapted well.

-Jennifer H., E-3, USAF
Intelligence/Cyber Specialist and Mother of one

I have been very fortunate and have not had to deploy or go TDY since my children were born. I am in a career field that doesn't require a lot of travel.

-Chelsea D., E-7, USAF
Medical Lab Technician and Mother of three

As a sailor, I have been away from my family for up to two years, which was awful. When I joined, I was under the impression that I would be with my kids more often, but I had to go to sea for extended periods. And you don't get a lot of phone calls or internet access. It was hard, but I knew it was the best decision for my family. I did this with three children and saw it as setting an example to show them they can do anything.

I felt like it was a strain for my kids because they could no longer see and talk to me every day. They asked why I left them, which was hard to hear. It was challenging, and I just tried to keep it light and think about their age and understanding of the situation.

-Mykeia L., E-3, USN
Logistics Specialist and Mother of five

My only extended separation was for nine weeks. It was good for me, I was busy. But, for my family, it was the first time I was away for that long. I feel like it was harder on my husband than on my kids. For me, it was when we talked, or video chatted that I realized how much I missed them.

For my kids, they had their times when they would break down, but it is crazy how resilient they are. They do so well.

-Kelsey Q., O-2, USAF
Operating Room Nurse and Mother of two

It was heartbreaking, and truth be told, you are never ready. I would start stressing as soon as I received notification of an upcoming deployment. I remember my youngest learned to walk while I was deployed and didn't remember me when I got back; that was hard, but that is the reality of what we do. I would also work to prepare my children and give them something to focus on other than my going away. I would explain it to them, and when they were younger, I would tell my oldest to take care of his brother and my youngest to take care of Daddy while I was gone. It gave them something else to focus on.

But deployments are not everyday life in the military. Sometimes folks get their information from folks who don't know what the military is about. People go on to retire, start that next career, and be very successful.

Because I had my boys later in my career, they were so young when I deployed that they don't remember. Of course, my husband and I remember much more, but they were so young.

-Cynthia S., O-4, USA (Ret)
Nurse and Mother of two

When I had to travel for work when my daughter was a baby, I did not feel horrible leaving her with the nanny. And sometimes, I took her with me and would find someone at those locations to come to the hotel to watch her while I was working. I can't believe how many sitters she has had. When I think about it, I can't even believe how much effort it was to travel with her. But, of course, it is easier now that she is older, and my husband and I are together at the same location.

-Melissa B., O-8, US Coast Guard
Chief Counsel and Judge Advocate General and Mother of one

My son is now in school, and I think it is one of those things that is a little counterintuitive. When they are young, it is not as disruptive, but it matters more once they are in school. I don't think that my son even remembers my absences. And luckily, I have not had to be away extensively or have not had to deploy since my son was born, but I think it goes back to finding ways to stay in touch and having as much contact as possible. Like using FaceTime to say good night, hear about their day, and just check in, even when not physically there. My husband and I are both in uniform, so I think I faced this experience as a military spouse when my husband deployed twice after our son was born. Trying to keep the family together as much as possible when scattered all around, you do what you can to keep the family feeling close.

-Janell H., O-3, USMC Reserves
Logistics Officer and Mother of one

I had to go TDY a lot as an active-duty mom and continue to go as a Reservist. I didn't go TDY until my first child was six months old, but I am gone about every month or so since. And I was breastfeeding and had to learn how to make it work. I even had to give up one TDY because I was breastfeeding, and it wasn't going to work out. That was the hardest thing, trying to figure out how to express and preserve my milk while TDY and traveling. And later to share that with other servicewomen trying to do the same thing.

For me, going on these TDYs was nice to have a different purpose. Being in the military is a higher calling, and I think it was good for me, but it was also a challenge to reconcile my roles as a military leader and a mother. Until I had children, I could put 100% into my role as a servicewoman, and my husband understood that because he was doing the same thing. But, becoming a mom changed that.

-Corey G., O-3, USMC Reserves
Combat Engineer Officer and Mother of two

I have not deployed since my children were born, but I have overseas exercises that require me to regularly be away from the family. When I had to finish up my training at Quantico and be away from my oldest son for two weeks, I had my youngest with me because I was breastfeeding. So, I was living in a hotel and had a friend come and stay with me to watch my son while I was working. But my oldest son was home with his dad, and my husband would tell me that he wanted his mommy back. That was the most challenging part...the

75

impact on my son at that moment.

-Maria F., O-3, USMC
Training Officer and Mother of two

Fortunately, my husband nor I have deployed since we had our children. I can't imagine how hard it would be to leave the kids behind. We both deployed earlier in our careers; we actually met on a deployment. My husband still goes TDY quite often, and I have sometimes felt like a single parent. For example, I had to PCS from Virginia to California with a small child while pregnant while my husband was in training to become a Chief Warrant Officer.

I have had to go TDY for about a week or so at a time, and I worry about them while I am gone. My husband is a great dad, but I still worry when I am gone.

My children are still young, and it is all they know.

-Jennifer Port, E-8, USMC
Administrative Officer and Mother of two

I have not deployed, but I have traveled a lot for work. There have been jobs when I am gone as much as I am home. Sometimes, I work fourteen-hour days; At one time, I called myself deployed to the Pentagon. Those times are hard. For me, I always try to make sure my kids are covered. I spent a lot of money on childcare, but I didn't care. I wanted to make sure I could find someone to take my kids where they needed to be because it was their joy in life. It takes planning.

But equally, right now, I am in a job where I can be there more. And I will rush home to get to the soccer field to pick them up, and they don't want to go home. They are happy and productive where they are. So, I question myself about who I am doing it for. Am I projecting something on them that isn't there for them?

Modern technology really helps right now as well. We can keep in touch and communicate when I have to be gone for a week or two. I keep it going for my children. I prepare meals ahead of time, I make sure they have their comfort items and set things up so they have fun and it's not like, "Oh, Mom is not here," but rather, "Mom is working, but we can still go on with our lives."

I think there are pluses and minuses for my children. I think there were times when I am sure they would have loved to curl up and snuggle with me. But there were other times when they completely and utterly surprised me and grew. One time, I was TDY, and I came back, and my oldest son told me he did laundry. I asked him to show me what he did. Turns out he didn't use soap. But he thought he had a super week, which proved to me that he could do more than I allowed him to do. Sometimes, it allowed them to grow in ways they may not have if I were around because I would have done the "mom" stuff. I would have done things for them.

-Rojan R., O-5, USAF
Program Manager and Mother of three

CHAPTER 7: IMPACT OF MOM'S SERVICE ON A CHILD

When one member of the family joins the military, the entire family does.

-Author unknown

As servicewomen, we have willingly joined the Armed Forces and accepted the responsibilities of that service. Likewise, as mothers, we are willing to take on the responsibility of raising our children to be strong, productive citizens of the world.

Being a child of a service member is as unique as being a service member. They are known to many as "military brats," not as a negative moniker, but to identify them as children of military members and, I think, to indicate their uniqueness.

Their lifestyle can be different from most children's. Often, they grow up away from extended family members and in other parts of the country and even various parts of the world, sometimes moving every few years. This means that our children also serve in their own way and can feel the impact without choosing to be a part of the military. Instead, they must navigate life in the Armed Forces and deal with its implications that we, their mothers, may not always understand.

As moms in the military, we must work to help our military brats adjust to the many changes that come with growing up in this military lifestyle. It is also our job to ensure they have the tools to understand our role as service members, why we serve in the Armed Forces, and how they fit in as military brats.

Nowadays, even the military does its part to assist our children in adapting to life as military children. The military uses the month of April as the "Month of the Military Child" to recognize their sacrifice and contributions. In addition, most military installations provide outlets for military children. Everything from a place to hang out to opportunities to participate in organized sports to after-school activities and, in some cases, a school to attend. A school where they can get to know and grow with other children who share their unique experiences. Many children with parents in the military must endure moving every few years, and with each move comes a new school, a new set of friends, a new community, new opportunities, and new challenges.

But being a military brat is not all bad. If it were, I think fewer of us would take on the challenge of parenthood and service. Growing up in the military offers a different lifestyle; for every negative, there is a positive.

Our children often get to experience the various cultures across the United States and around the world. Experiences that come with their status as military brats. Their level of independence, resilience, willingness to experience new things, and self-confidence often blossoms ahead of their contemporaries. Life in the military and all the changes they experience include relationships with other children living the same lifestyle. They can and often do develop lifelong friendships with people around the world that they can carry to adulthood, all made easier by today's technology.

We hope that the impact on our children is more positive than negative, and we do all we can to ensure that their lives run as smoothly as possible, even with all the changes that come with military lifestyle.

As a mother of military children, I sometimes worried about the affect military life had on each of them. They each faced their own struggles and adjusted in their own way. I often saw the pain or frustration that life in the military can bring to a child. On more than one occasion, I saw the struggle to fit in with family back home, to catch up and keep up with family relationships. Sometimes their battle was just moving to that next duty station. I remember one move when our oldest child desperately wanted to stay with her best friend. Or when my other child's best friend moved away, being only six, she could not really understand why her friend wasn't going to be at school anymore. But the frustrations experienced with moving often became an asset when learning to meet new people and experience situations. There was so much positive impact as well. My children were exposed to several different cultures that I hope expanded their appreciation and understanding of the world around them. Life in Arizona was undoubtedly different from life in Texas or South Carolina. And life abroad certainly helped them appreciate life in the United States and be open to new cultures, people, and new experiences.

-Patricia Q., E-7, USAF (Ret)
First Sergeant and Mother of three

Even though my son was young, even when I got out of the military, we often spoke of my time in the Navy, and he still remembers those days. In the Navy, I experienced a lot of firsts, and my son was always with me. He remembers being young and around the military, and especially the aircraft. I think my experience was a part of why he later joined the military himself. His dad and I both served, and he followed in his dad's footsteps and joined the Marine Corps.

-Jacqueline G., E-3, USN Veteran
Personnel Specialist and Mother of two

I like to think the military lifestyle has been positive because I make it a point to focus on the positive. I know my oldest feels proud seeing me do something I enjoy. My girls know how much I enjoy what I do day to day and seeing that has had a positive impact on me.

-Anna P., E-5, USAF
Services Specialist and Mother of two

When my daughter was growing up, having been born after I joined the Air Force, it was the only way of life she knew. She had nothing to compare it with. I think the experience has helped her be more resilient, independent, adaptable, and able to accept change. However, I feel that because of the sacrifices she saw me make as a single parent, she does not have a desire to serve in uniform. She told me she would prefer to work two part-time jobs rather than join the military. I also think that she does not want to make such sacrifices again because of the sacrifices she had to make growing up in the Air Force.

-Pagerine J., O-5, USAF (Ret)
Medical Service Corps and Mother of one

My husband and I have twelve children, ranging in age from two to twenty-five, and we have always tried to get buy-in from our children when moving from post to post. We made it a point to always ask them how they were doing and felt. There are times when I know that we've discussed it, and they wanted to know when I would retire. Then COVID-19 hit, allowing us to see what that would be like.

When we lived in Texas for eight years, my husband was doing the coming and going while we spent those years in one location. It was here that my children had more stability and made a life for themselves. So leaving Texas, for them, was a bit challenging. But I tried to find and highlight the silver lining. Again, not minimizing their feelings but maximizing the opportunities ahead of them and helping them move forward.

-Leslie L., O-5, USA
Human Resource Officer and Mother of twelve

Overall, I think it was a positive experience. And, like I said earlier, I think it made my son more resilient. I know we throw that word around a lot, but it's true. I believe our military lifestyle has been a great advantage for him compared to those who remain in one place their entire lives. Because he has traveled and met so many people, he is a very accepting human being, and I appreciate that. For example, I remember being in Germany, and he was able to play with the German children without speaking a word of German. So while we could have taught him these things, his ability to experience them for himself is quite different.

-Yulanda B., O-4, USAF (Ret)
Personnel Officer and Mother of one

When I retired, my boys were so excited about it, and they told everyone in their circle that I was retiring from the Army. They were excited, knowing that I was going to be around more.

In fact, shortly afterward retiring, I remember taking a call about volunteering, and my youngest was so upset because he thought I was going back to work. So, I learned, after it was all said and done, about the impact my service had on them. That they missed me more than I knew.

-Eboni B., E-8, USA (Ret)
Career Counselor and Mother of two

As they got older, my children did miss out on seeing family. But in the end, my service, our military lifestyle, and their exposure have made them more flexible, adaptable, and resilient in any situation. That is because of their exposure to life in the military.

I invested a lot in them so that they would be able to make strong decisions. Today, they are strong, compassionate, well-rounded, college-educated individuals. Even after my service, they have continued to travel around the world on their own, enjoying their new experiences.

-Patricia H., O-5, USA (Ret)
Judge Advocate and Mother of three

I think my service does impact my children. But I think it is primarily positive. I have not had any traumatizing experiences that would affect them, so overall positive. I talk to my children about my service and why it's so important that we have a military to help them better understand. I also try to instill in them a sense of determination, hard work, and how it pays off.

My oldest just graduated kindergarten, and I was so proud to see her get the award for being the hardest working student. So, I think she sees me working hard, and she is willing to do that.

Also, change is not a bad thing. The move was hard on my husband, who is not military. And it took a while for us to adjust to the move, which helps our children see that change can be good. And we can see and show them that sense of resiliency.

The military has also helped my children see the camaraderie that comes with life in the military.

-Toni D., E-8, USAF
Personnel Specialist and Mother of three

I think that when our children become adults, they will understand and appreciate the impact of the military lifestyle. Growing up, my mom was a reservist, and it wasn't until I was an adult that I could understand the positive influence of her service and the sacrifice she made by being away when duty called.

-Ebony A., USAF Spouse
Mother of three

Overall, my children have adjusted well to my service. Especially the separation due to my TDYs, the military required moves, and going to a new school so frequently. My service really had some consequences for my oldest really; she did develop separation anxiety with me being gone so much and spending so much time away from me and in daycare even when I was home. But I think my time in the service also instilled a sense of strength and determination in my children.

-Kimberly B., E-5, USA Veteran
Logistics Technician and Mother of four

My daughter is young right now; she is only four years old. She has already been meeting new people and experiencing new things. So far, this has positively impacted her, even her just seeing me in my uniform. I think it lets her see me doing something meaningful, and I hope it helps her know that she can do whatever she wants.

-Jennifer H., E-3, USAF
Intelligence/Cyber Specialist and Mother of one

My children are so young that I don't think they know or understand what I do. They are young now, but if I make CMSgt, I can stay for thirty years (as is my plan). Of course, by then, my oldest will be eighteen, so we will see. But right now, I think the impact is positive.

-Chelsea D., E-7, USAF
Medical Lab Technician and Mother of three

I do think it has had an impact on my children, a positive one. At first, I thought it was a negative one, but I have realized that it hasn't been all that bad. Maybe because I am with them now. Recently at Thanksgiving, my children, while doing a school project, helped me realize they are proud of me. My oldest said that she felt I was a superhero because I put the needs of others before myself, and my younger sons were proud of me as a "vet" or "veterinarian," but I knew what they meant. I questioned if what I was doing was right, but after that, I realized and now know that they appreciate my service in their own way.

-Mykeia L., E-3, USN
Logistics Specialist and Mother of five

So far, the impact has probably just been adjusting to our new lifestyle, putting themselves out there, and meeting new people. My son is more outgoing, while my daughter is more reserved. I think they will make friends and meet people they would not have otherwise met. In addition, they are connecting with other military children who can share and understand what they are going through. When I think about the last year, our first year in the military, we did a lot, and if anything, the next move will be easier.

-Kelsey Q., O-2, USAF
Operating Room Nurse and Mother of two

My children loved the traveling we got to do; they can tell you all the places we lived. Even now, at twelve and thirteen, I think life in the military has helped them. They are social butterflies; they can go anywhere, get acclimated to the situation, and make friends easily. I think that comes from growing up in the military. The military puts its members altogether, gives us the mission, and makes us work together. Like us, our children have learned to get along with people who don't look like them, or that might be different from them.

-Cynthia S., O-4, USA (Ret)
Nurse and Mother of two

My daughter is still young. She hasn't been able to see my parents for quite some time due to COVID-19. But my sister, who has been vaccinated, comes down to visit, which has been good for my daughter. She is an only child and a child of "old" people. So she thinks that my friends are her friends. But it's good, and she's good. She is not scared of adults and even thinks of them as her peers.

-Melissa B., O-8, US Coast Guard
Chief Counsel and Judge Advocate General and Mother of one

I think my son is young and military life is all he knows. But, he seems to be proud of what he can recognize. I think the other piece is just meeting others with similar experiences of growing up in the military and being a part of this community where we put our lives on the line for our country. Regardless of what branch you serve in, just being exposed to other military people is a huge reward. And the families and friends we have met are great people that I want my son to grow up around.

-Janell H., O-3, USMC Reserves
Logistics Office and Mother of one

I grew up as a military child, which is almost a point of pride for me. I didn't grow up in the same place my whole life. I experienced different things and I have interesting stories. My husband is the opposite; he lived in the same house all through high school. I think that offers a certain appeal, but my parents instilled a strong sense of belonging in our family, and we were it. We were strong together; it was about family, not where we lived. I have tried to carry that over to my children. Our home is our family. It's been great for building resiliency.

My oldest child is only three and so empathic, and the connections he has been able to make are great; it's really cool to have conversations with him. He's pretty perceptive, and he does act out when things change.

In some ways, it is positive because they see me do something I am excited about and get to tell them about it. And in some ways, it is so hard for them. Like when I have to leave, they get mad; they are too young to get it. And when I get back, it is an adjustment for them. I try not to take it personally, but it does hurt. Fortunately, my husband is great and helps me get through those times and tells me how great it will be for me to tell my grandchildren what a bad-ass I was serving in the Marine Corps.

-Corey G., O-3, USMC Reserves
Combat Engineer Officer and Mother of two

Because my children are so young, I am not sure there has yet been an impact. But speaking optimistically, there is not a massive amount of negative impact. We explain things in a way they can understand and provide the bigger picture of what we are doing by serving in the military. Whether or not they understand it at their age, we don't know. They really just know what is happening at the moment, which can be challenging for them.

We are coming up on our overseas control date, and one of the significant issues for us is how do we manage that, and can we reset that date? Do I do a one-year unaccompanied assignment, or do I do a deployment to reset the clock? Unfortunately, for medical reasons, I cannot take my family overseas, so this is something we are dealing with and weighing out the options for us. We are also dealing with the possible consequences these decisions will have on our children. So, it's definitely not an easy choice.

-Maria F., O-3, USMC
Training Officer and Mother of two

I would say the impact of my service has been positive on my children. They are adaptable and very "go with the flow." We are going here and there, and they just go with it. My kids are six and seven and have traveled more than most adults in the U.S. They have been all over the place. They are up for anything, their life is an adventure, and they love it. An excellent experience for them.

This is the second year our kids have spent time with their grandparents and extended family during the summers. We are very fortunate that our parents can take them on and give them that time. The opportunity to get to know their extended family is excellent.

It's also been great for us. The house is clean, and I get a lot of me time. We eat what we want and have date nights. It's great. It's nice that they spend time with family, but it's always great when they come home.

-Jennifer P., E-8, USMC
Administrative Specialist and Mother of two

It is not necessarily an easy life, picking up and moving. Fortunately, for my kids, I have figured out a way to not have to move so often. But still, there is a certain amount of resiliency, a certain amount of putting yourself out there, and a certain amount of learning and growing required.

Humans are resilient, and I think military children are even a little more than the regular population. I think sometimes we worry too much. Our kids are going to be okay, and we are going to be okay. I think as long as things are done with love, understanding, and a "we can do this" attitude, there is something about your mental state that makes it work, and I can see it in my life.

I hope my service has given them a higher calling, a reason to serve, and a better appreciation of the world and the things around them, the level of appreciation for the things we, as Americans, have access to. I remember seeing a video of one of my sons reading a poem he wrote about Veterans. I think that's the benefit; they will see the world and the people around them differently and maybe see America a little differently. Sometimes it's big pie-in-the-sky stuff, but I think it's true that they can appreciate and see the world differently because of the lives they are living.

-Rojan R., O-6, USAF
Program Manager and Mother of three

CHAPTER 8:
FEARS AND/OR REGRETS OF SERVING AS A MOM

Once you start working on something, don't be afraid of failure and don't abandon it. People who work sincerely are the happiest.[15]

-Chanakya

No matter what we do in life, we are bound to have fears, and no matter how great we are doing, we may experience some regrets. This is universal and inescapable, and whether you serve in the Armed Forces or not, your world can be full of "would have…should have…could have."

For some servicewomen who have children, one of the biggest fears can be whether life in the military suits their children. Especially when it is time to pack up and head to a new location, leave on an extended TDY or deployment, or when your child is upset about leaving friends and schools that they have come to love. Like other moms, we sometimes experience fear about the decisions we make and the impact these decisions have on our lives, our spouses, and the lives of our children, not just in the short term but in the long term as well. For other servicewomen, the fear can be failing at service, motherhood, or both. When faced with the unique challenges of serving in the Armed Forces and motherhood, the worry of doing both and doing both well can weigh heavily on you. Sometimes, the fear is about the effect of our service on our children. Are we doing the right thing? Will they grow up and resent the lifestyle we chose for them? For most of us, these fears are short-lived as we realize that we have made the best choices for ourselves and our families. I believe that making the best choice for yourself (and, by extension, your family) is almost always the best choice for those around you. If you are not doing what is best for you, you are not your best self and, by extension, not the best you for your family.

For some of us, there might be regrets. Regrets that we must face and accept. In the end, regrets, like fear, are a part of life, and serving in the Armed Forces does not exempt us from them. And like everyone else, we hope they are small and inconsequential compared to the our chosen path and the many rewards it brings us…both as servicewomen and moms.

While life can be full of fears and regrets, we do our best to not let our fears make decisions for us; and we try to make sound decisions so that when looking back, we don't see a life full of regrets.

When it came to my duty as a servicewoman, I don't know if I experienced a lot of fears about my service. I sometimes feared that my role as a mother was incompatible with my role as a servicewoman. But I found that when I took care of work, there were times when I could leave early or take a long lunch to meet the needs of my children; and if I took care of and focused on my children when at home, there were times when I could take some extra time to focus on work. Additionally, I was blessed to have a husband who would take up the slack when my duty took me away from home or meant not being there for our children. So when I think of regrets, I don't have many, but like most of us, I have had a few. When it came to my service, I had no regrets but sometimes felt some uncertainty about raising my children in the military. Because they didn't get to experience extended family in the same way that I did or get to know my parents as well as they would have if we lived closer. Especially when I thought there would be time for them to get to know family later, and shortly after retiring, my father passed away. But when I ask myself if I would do it differently, I have to say no. I always felt my service to my country was a calling, and I believe it was best for me. And I think I took comfort in knowing that what is best for me, as the parent, could not be all bad for the children.

-Patricia Q., E-7, USAF (Ret)
First Sergeant and Mother of three

My only regret about my service was the decision to get out, not because I wanted to but because my husband wanted me to. He knew I was in the military when he met and married me, but he didn't like the idea of me being in and traveling, so I decided to get out. I think otherwise, I would have stayed in. I think there was still some growth that I could have gained and some more training I could have obtained.

-Jacqueline G., E-3, USN Veteran
Personnel Specialist and Mother of two

I think the only fear I had was right after having my children. Knowing I would have to deploy, I feared they would not remember me after deployments. But, of course, they obviously remembered me. And when I was gone, we could FaceTime and keep in touch. So that fear quickly went away.
As for regrets, I don't have any because there is no point in focusing on regrets. I am just not regretful for joining the military.

-Anna P., E-5, USAF
Services Specialist and Mother of two

I think I didn't have any fears as a spouse. I guess my biggest fear was for my son's health, and God has blessed us. It would have been different if we moved every few years; finding schools and helping our son adjust to the new locations would have been hard. In some ways, it was like being married to a civilian.

I don't have any regrets because the military has been so good to us in caring for our son. Maybe one regret was not living closer to family to have their help and support or having the opportunity to experience living in different locations. But it's so tiny when I look at what we have here in Texas.

-Carol G., USAF Spouse
Teacher and Mother of one

Fears of not coming home from deployments and how that would have impacted my child. Fear of not being there and missing milestones and special events in my daughter's life.

Regrets: the regrets of missing out on those special moments and milestones and mainly having caused my daughter to worry about me.

-Pagerine J., O-5, USAF (Ret)
Medical Service Corps and Mother of one

When I was a lieutenant, my husband said, "We don't see you." And I had to get control over my time. As I become more senior and my responsibilities increase, one of my fears is that I will go back to being a workaholic. So, I am very conscious of that, and I try to make sure my kids understand my role and position. For instance, when I knew I was going to the Pentagon for a few months, I knew I would be working long hours and that my time at home would be limited. So, I prepped the family, letting them know it would be a temporary thing. And I maximized the time we did have together.

I don't have a single regret, and I am now in a position where I can retire whenever I want, but I choose to continue to serve.

-Leslie L., O-5, USA
Human Resource Officer and Mother of twelve

My greatest fear, other than going to war and dying, was my son feeling that I was not as supportive as I could have been. I wanted him to know that even though I could not always attend his functions, he had my support. And the fear of us being divided rather than close, that I couldn't understand what he was going through. I wanted him to always know how much I loved him.

I have one regret. On my son's first day of school, I still get upset thinking about it. I was in training, and my supervisor refused to let me step out for a bit to see him off to his first day of school. At that moment, I felt like I put the Air Force before my son. After that, I told myself I would never allow something like that to happen again. That is my only regret.

-Yulanda B., O-4, USAF (Ret)
Personnel Officer and Mother of one

I worried that I had broken my children. In my absence (deployments), my husband had set everything up like a well-oiled machine. And my coming back and trying to assert myself and what I was doing was contrary to what he had set up, and I was doing it wrong. He had set a standard, and I wanted to come back and get involved; I felt like I broke it. I remember when my first deployment was changed from twelve to fifteen months; it was only ninety days, but it meant I would miss two birthdays, making me sick to my stomach. There was so much anxiety about missing those two birthdays. It's doable, but it is hard.

But I don't have any regrets; I don't. I feel like I am retired at this point. One of my most significant accomplishments was having my children see the world while they were old enough to experience and remember it and then putting them in a community where they are completing high school. I honestly feel like my boys had the best of both worlds.

-Eboni B., E-8, USA (Ret)
Career Counselor and Mother of two

My biggest fear was not being an adequate mom, and that I wasn't going to be able to teach them everything they needed to know. But I determined early on that I would not miss out on anything I didn't absolutely have to miss. So I worked hard to really be there for my children. Even if that meant leaving the office early sometimes to be there for them.

93

For regrets, I regret not giving my children responsibility for things that were theirs. Doing something that I should have had them do just because it was easier for me. That meant that I did a whole lot more than I should have. I didn't have a bigger plan of how I could enrich them without just making it easy for myself, like getting takeout when I could have been teaching them to cook.

Both times I deployed, I came home to find out that I was not getting promoted, and at those times, I wondered why I went through all of that if it wasn't going to help me professionally; especially knowing that part of the reason was that I was a Black woman. So, yes, those glass ceilings were still there.

-Patricia H., O-5, USA, (Ret)
Judge Advocate and Mother of three

I definitely worry about missing out on things in my kids' lives. I think that is my biggest fear...just not being there. I was stationed overseas for a long time, and now that I am back, I know that I will have to deploy sooner or later, which will be hard. But, I worry about not being there for a birthday or a holiday and that my children will remember that I missed a milestone. Even if it's a sickness or injury. I worry that they will remember that I was not there for them.

I can say that I don't think I have any regrets, at all, about being in the military.

-Toni D., E-8, USAF
Personnel Specialist and Mother of three

As a military spouse and mom, I have to be honest and say I didn't want my children to join the military. Not that it wasn't a good experience, but I saw how challenging and difficult military life can be; so that they would join the Armed Forces was a fear for me. Also, I feared the impact military life had on my children, all the moving, having to make new friends, and fitting in. In addition, my children are bi-racial, so there was always a concern with each move. Depending on where we were stationed, I wondered how we would be accepted.

I always regretted leaving the friends I made, and I am so grateful for today's technology. It has allowed us to stay in touch. That, and the

fact that I never got to have a career as a military spouse. With his absences and the constant moving, I never got the opportunity to have a career/business. I also put my educational goals on the back burner.
-Ebony A., USAF Spouse
Mother of three

One fear was realized when I came home from that final remote assignment, and my daughter didn't recognize me. Another was being judged, as a Black female, by co-workers. Also, as a soldier, you are expected to suck it up and drive on. Ultimately, a fear of failing, failing myself and my children. Fear of being unable to accomplish things and support my family as a single parent.

But I don't have any regrets. Because I loved the experience of being in the military and loved being a single mom.
-Kimberly B., E-5, USA Veteran
Logistics Technician and Mother of four

I guess my biggest fear is how often I will have to deploy. One of the reasons I chose the Air Force was that their deployments were shorter. My other worry would be being unable to see my daughter while working my shifts. But, so far, it has not been too hard. It really is working well for both my daughter and me.

I have no regrets at all, not one. On the contrary, I enjoy my job and the peace of knowing that I am securely employed, and that healthcare is available. Especially as a single parent, I am glad to be serving.
-Jennifer H., E-3, USAF
Intelligence/Cyber Specialist and Mother of one

You know what, it's always one of those worries that if we have a situation where things are so severe, and restrictions are in place, and I am stuck in the hospital and unable to go home. Being close to my children but unable to get to them would be horrible. For me, even more so than being across the world. We came very close to that during the pandemic. Since I work in a hospital, it was a real possibility.

I don't have any regrets yet. But I would love to do a short tour, and if I don't get to do that, I will regret not having the opportunity to have that experience.

-Chelsea D., E-7, USAF
Medical Lab Technician and Mother of three

My biggest fear is my rate (pay grade) and having to be separated from my children as they get older. I will have to go back out to sea or go on deployments. With my husband in the Air Force and me in the Navy, my husband and kids may not be close to where I would be stationed. Meaning I would have to be away from them. Also, the fear of something happening to me.

The only regret I have is that I didn't do it sooner. Meaning joining the Navy, making rank, and having my career path set before I had children.

-Mykeia L., E-3, USN
Logistics Specialist and Mother of five

So far, my biggest fear is that my kids might resent all the moving. My husband and I have a group of friends we grew up with, and I worry that my kids won't have that core group of friends. But then I realize that a lot of people don't have that. I know they will find their core group wherever they go. I felt like I was robbing them of that experience, but at the same time, I think that they will be getting a different experience that will "better" them.

-Kelsey Q., O-2, USAF
Operating Room Nurse and Mother of two

You know what? You always have a fear of not coming home. Even though it could happen at any time, being in the military puts us in harm's way (and thank God, my husband would help keep me on track about it). That, and just being away from my kids.

I do not have any regrets. The truth is that the military has provided so much for my family and me that I have no regrets.

-Cynthia S., O-4, USA (Ret)
Nurse and Mother of two

I am in the Coast Guard and more senior in rank, so, at this point, my biggest fear is getting a paper cut (laughing). But, no, I don't have any fears. Although, I don't take the risks I used to take when I was single. A little less risky. I think if I was deploying, I am sure I would feel differently.

No regrets. This has been the best part of my career; it's just gotten better and better, and it's the best job for me. I never dreamed I would have gotten it, and I love it and the people and our community. Had I not gone to law school, I might have thought of other options, but I have no regrets.
-Melissa B., O-8, US Coast Guard
Chief Counsel and Judge Advocate General and Mother of one

This question was kind of hard when I was thinking about it. Part of it, for me, was that I had a real burnout experience. I had gotten to the point where I had broken myself. I think it was from an overly unrealistic expectation and sense of perfection for myself; you know, trying to be the best Marine and mom. I have gained a much better perspective over the last two years. So, I think one of my regrets was not having that balance sooner. It was at a point where my days were so detailed down to the minute that I didn't leave myself any wiggle room. And since you can't predict everything, the unexpected things would throw me off. I think that was my fear and my regret. Fear that I could not do it and regret that I didn't realize earlier that I needed a more balanced life.
-Janell H., O-3, USMC Reserves
Logistics Officer and Mother of one

The usual fears that any new mom would have.

When it comes to regrets, "regret" sounds like a really strong word but I didn't necessarily want to leave active-duty. In the end, when we talked about, it was the right decision for my family. My husband was getting out and the Reserves was a way for me to stay connected to the Marines and have a more flexible role.

-Corey G., O-3, USMC Reserves
Combat Engineer Officer and Mother of two

I don't know about fears; I guess I haven't had a whole of time to worry about that. I try to be very optimistic, but I guess the biggest fear is being torn away from my family because of the military. Fear of not having the patience and communication to understand what my family is going through, letting work consume me in the moment and not thinking about the stressors that my family might be experiencing. I think the key is coming home and asking the questions first rather than bringing my work life home and not know what they are going through. Also, fear that I will forget how to do that effectively and drive my family members away. It is hard to make that a tangible fear but fear of losing my family emotionally.

-Maria F., O-3, USMC
Training Officer and Mother of two

Fears? Let's see…so it's more of a future fear that my boys will join the Marines. And how do we tell them to go to school or take a different path when both their parents served? I do fear that maybe I will see that I wasn't there enough.

Regrets? I don't think I have any, really none.

-Jennifer P., E-8, USMC
Administrative Specialist and Mother of two

I think every mom, serving on active duty or not, worries about messing up their kids, or at least I do. My goal is to not mess them up, and that's really what I worry about. I don't worry about their health or safety so much. I fear that they will not feel loved enough or that they can talk about how they feel. Suicide is big, and I want my kids to know they are loved and can always come to me. I am raising Black children, and I have to talk to them about how life is not fair. They have to be intelligent and knowledgeable about the world around them. I think I have the same fears as other moms. The same concerns of sending them out in the world and having their hearts crushed. I am the same as every other mom; I am just a mom who wears combat boots to work.

I would say some of my regrets are not saying yes when I should have, and I wrote about that in my book. I think I held myself back from what I could have accomplished. I said no because I could not see how to do it. I couldn't see how to make it all work at that moment. I should have said yes; I should have been more possibility-minded in my ability and my kid's ability, and my family's abilities. I said no, but I want other women to say yes.

-Rojan R., O-6, USAF
Program Manager and Mother of three

CHAPTER 9:
REWARDS AND
MOTIVATIONS OF
SERVING AS A MOM

We need to reshape our own
perception of how we view ourselves.
We have to step up as women and
take the lead.[16]

-Beyoncé

I am not sure who said it, but there is an old saying, "Anything worth doing is worth doing well," and another one that says, "With a job well done comes many rewards." Even today, these clichés are valid and can apply to any and everything we do.

It is the same in the Armed Forces, maybe even more so. The military must operate as a well-oiled machine and requires each member to be ready and able to do the job they are trained to do. Not just for us but for our family's sake, our fellow service members, our nation, and our allies. This creates a feeling of knowing what we do every single day matters, not just to us but to our family, neighbors, and our nation.

I am sure that very few believe that serving in the Armed Forces will bring you financial riches, but the rewards are many and can be as unique as service in the Armed Forces. It can be the fulfillment of a desire to serve your nation and make a better life (for yourself and your family); for many, this reward keeps us in the Armed Forces. It can keep us putting on the uniform and moving forward to that next assignment, deployment, and challenge.

Other rewards can include education, a skill set for the next chapter of your life, medical benefits for yourself and your family, confidence in who you are and what you are capable of, leadership skills, and the opportunity to see and develop friendships worldwide. They can also include, for some, a retirement income, healthcare for life, and even education for your children. These same rewards also empower us to keep doing all we can and allow us to create the best environment for our children today and their future opportunities.

Serving in the Armed Forces as a mom can also provide rewards. Combining service and motherhood can also bring unique rewards for each of us, our children, and the generations to follow. These rewards can also be the motivation that keeps us in uniform. For example, it could be the reward of showing our children the possibilities of service in uniform. Or to show our children, and the world, what women are capable of, or the level of success women can achieve in the Armed Forces. Many consider it a reward to have the opportunity to be a voice for women everywhere while making room for women throughout the Armed Forces, especially in our senior noncommissioned and commissioned ranks.

The rewards and motivation may differ from woman to woman, but what is true for all of us is that we proudly serve (or have served), even if how or why we choose to serve in the military is different. Also true for each of us is that our service in uniform comes with many rewards.

When I look back on my service in the military, many rewards came with being a servicewoman and a mother. I held four jobs in the military, and each was rewarding for me. First, as a mechanic working on flight line equipment, this helped me understand the bigger Air Force mission. Second, as a military training manager, I was entrusted with the military training for our newest recruits. Seeing them become less civilian and more military meant I was again part of the bigger picture and contributing to the Air Force as a whole. Third, as a paralegal, I worked with attorneys, judges, and base leadership to ensure that the military justice system was working, not just punishing those who violated the law but protecting those who needed it. And finally, as a first sergeant and part of the leadership team, I got to use my experiences and leadership skills to work with and mentor my fellow airmen. All these career choices brought me rewards in their own way. And, serving as a mother, I found that raising my children helped me see the world from a different perspective. It helped me be a better Airman and a better leader. Motherhood helped me to see that the world was so much bigger than me and made me want to create a better world, and, in that same vein, it helped my military career.

-Patricia Q., E-7, USAF (Ret)
First Sergeant and Mother of three

I think that there have been so many rewards. I have been able to travel. Also, with my troops, they can be like my military kids, and seeing them grow and develop has been very rewarding.

What keeps me in the Air Force? It was always my initial goal to have a military career, and for my children, I wanted them to see that joining the military is something you can do whether you are a man or a woman. My daughter always comments on how many men are in the military, and I wanted her to understand that women can do anything they want, including the Armed Forces.

-Anna P., E-5, USAF
Services Specialist and Mother of two

The biggest reward of being in the military was, once again, having that medical care for our son. I also think because my husband served, there were no issues with him having to take time off to help take care of our son or attend medical appointments. So, that was a huge relief not having to do these things on my own.

And, I have met some really good people; looking at all the people I have met, you would think we have moved around as much as any military family.

-Carol G., USAF Spouse
Teacher and Mother of one

One of the rewards of my serving was showing my daughter that there are things bigger than yourself. It is not always about us as individuals but about making a difference. You can certainly have influence outside the military but serving in the military is a sacrifice. For example, in addition to your primary job in the military, we do many things that are unique to the military. I initially joined the Air Force for school, and I did get that, but I think my service has shown my daughter that there are things worth fighting for...things tied to freedom.

What kept me in the service? Well, to start, being able to keep food on the table, a roof over our heads, and medical care for myself and my daughter. But after that, it was because there were very few women O-6s, let alone women of color, in the medical field. So, I wanted other young Black women to see that it was possible, and to highlight the need for diversity in officer ranks in the Medical Service Corps.

-Pagerine J., O-5, USAF (Ret)
Medical Service Corps and Mother of one

Just yesterday, my son got an offer from West Point and the Naval Academy; it was amazing. And my son said to me, "Mom, I brag about you and Dad all the time. I have my two military parents serving, and that is a part of me and who I am." That's it. My kids are sticking their chests out and are proud of their parents and our impact in our little sector. So it was very cool to hear my seventeen-year-old say that.

Part of what makes me stay is that I want to be in the position to be a voice, to show others that you can be a mom, be successful, and advance in rank. Of course, you can do all that, but here is what it will take. I want to be authentic and say that you must work hard, plan, speak up, and sometimes exercise personal courage with your boss. I have done it, and I want to be able to empower others to do the same.

-Leslie L., O-5, USA
Human Resource Officer and Mother of twelve

Serving as a mom, my son has been able to travel and see so much. He has had the opportunity to travel to different countries, visit various sites, and expand his realm of friends and family. And he has established his own network of people around the world.

What kept me in? For me, if I start something, I always want to complete it, and I want to teach that lesson to my son. To finish what you start, no matter what society has to say, you do what you need to complete it, even if you have challenges along the way. I also wanted to ensure I had the financial security to take care of my son if something happened to my husband.

-Yulanda B., O-4, USAF (Ret)
Personnel Officer and Mother of one

I feel the sacrifices I made because of my service were all worth it. This includes benefits we didn't even think of, helping my boys get through college and buying a house. 100% worth it.

I would say what kept me in uniform was that serving in the Armed Forces was my "first love," and I knew my husband had my back. His support helped me realize that, at the end of the day, our kids would not lack anything during my absence. It was like the "golden ticket," and I was unstoppable, knowing my kids wouldn't miss a beat.

-Eboni B., E-8, USA (Ret)
Career Counselor and Mother of two

I am thankful that I learned that I could do more than one thing simultaneously. Growing up in the south, my mom was a stay-at-home mom. We had very little money and just did the basics. Then, I came into the service, successfully completed twenty-eight years of military service, and raised three children who went to college on the GI Bill (thank you, Jesus). I am pleased to see that I made an impact while in the service, and now, I continue to do the same in my post-service career. Today, I continue to mentor young men and women, hopefully making a positive impact.

I have much more wisdom, a much broader life, and a better quality of life than my parents had. I have a great life, and Jesus is a big part of it, giving me the proper perspective. So I ask myself, how do I help others the way Jesus helped me...and make a difference in someone else's life? So even after the military, I decided to work with the Vietnam Veterans of America so that I could continue to make a difference.

-Patricia H., O-5, USA, (Ret)
Judge Advocate and Mother of three

I think the reward for me in serving my country is showing my kids that you can do it (whatever it is) if you want to. Another reward is showing other women (and men) that you can do it even when other people might doubt your ability to do it.

What keeps me in as a mom? Honestly, I love the military; I have had many amazing experiences I can show and share with my children. It is also the compensation that comes with service. And, getting out now would mean starting over. But I look forward to spending more time with my children, and retiring from the military will allow me to do that. Besides, the military is all I have done. I joined the military when I was eighteen, and the thought of leaving prematurely is scary.

-Toni D., E-8, USAF
Personnel Specialist and Mother of three

Being able to raise my children as a military spouse, I am extremely proud of how well they adapt to changes such as having to regularly move. I have never been good with change and adjusting to each move would take a year to get comfortable again. But my kids, they just seem to roll with it. And I am grateful for the closeness my kids share. I think they learned the importance of family and to value it more.

-Ebony A., USAF Spouse
Mother of three

The greatest reward was spending time with my children. That was the best time; just spending time with them. I remember at Fort Carson, I had a sergeant major who really pushed family time and was serious about us not working late when we could be spending time with our family.

Another reward is just seeing my daughters blossom into the young women they have become.

What kept me in for as long as I served was the need to take care of my family. While I am no longer in uniform, I am still affiliated with the Army, and I continue to get rewards from my work with servicewomen and single parents.

-Kimberly B., E-5, USA Veteran
Logistics Technician and Mother of four

My reward is serving as a mom, which I see as being a strong role model for my daughter. Also, knowing that her future is secure, and that she knows that she has more options now that I am in the military.

I am not sure what my future holds right now. I am waiting to decide if I will continue serving in uniform. But really, just waiting to see what doors open for me.

-Jennifer H., E-3, USAF
Intelligence/Cyber Specialist and Mother of one

The rewards of serving as a mom. That's a good question. I am not really sure. But when I have a tough day, I can come home, see my kids, and get a hug and a kiss... that's the reward. I think another reward is the patience that I have developed. And, really, I have more of that "mama bear" mentality that I have used to work with and protect my airmen.

What keeps me in the service? My goal is to make CMSgt (E-9). Also, the benefits that my service provides. My family depends on me, and I, in turn, depend on my service to provide for them.

-Chelsea D., E-7, USAF
Medical Lab Technician and Mother of three

A lot of folks in the military are by themselves. But, at the end of the day, I get to go home to my children, talk with them, and know I am taking care of them. There is so much that I consider rewards for serving in the Navy.

If I make rank, I will definitely stay in. I am also submitting a package to become an officer. If either of these goes through, I will stay in. I am still thinking about what I will do if neither happens. At this point, I am not sure.

-Mykeia L., E-3, USN
Logistics Specialist and Mother of five

I feel like the rewards are endless for myself and my family. This has been great for all of us. When I think about if I will continue my service, I think so. I love what I do, and the most challenging part will be advancing my career and continuing to do what I love (working in the OR). I realize that I might have to do more administrative work as I advance in the military, which could mean giving up some of the work I love.

-Kelsey Q., O-2, USAF
Operating Room Nurse and Mother of two

I remember thinking about getting out of the military to attend nursing school, but the military allowed me to stay in and become a nurse at the same time. God just worked it out for me to get accepted into nursing school while serving in Korea.

In the military, you have to be focused; I did that as an enlisted member and officer. In or out of the military, life will pass you by if you don't have a plan and stay focused.

I was just coming into the service to get it together and stayed to retirement. When I did look at getting out and starting over, I could not see myself doing that. I was more than halfway to retirement. And again, the benefits and what I got from serving were so great.

-Cynthia S., O-4, USA (Ret)
Nurse and Mother of two

It gives me so much pride when I take my daughter to work events. Recently, I took her to a retirement ceremony where I was the presiding officer, and she was there in her miniature Coast Guard uniform. I was speaking, it was emotional, and it was fun. My friend told me that during the ceremony, my daughter said she was so proud of mommy, and I thought how great it was that she was proud, and I wanted her to believe that she could also grow up and be a leader. That makes me feel good. I think women who grow up with mothers who have careers tend to be more confident and have more career aspirations.

What keeps me in? I have one Coast Guard friend that I went to boot camp with. She is also a two-star Admiral, and we are still very close. We have a very unique voice, and our perspectives are unique. When we talk about things, it comes from a different place than our male counterparts. We feel very committed to the women of our service, which keeps me in. I think of the future generation and seeing the Coast Guard better than before, which is particularly important to me.

You know, there are not a lot of women in the Coast Guard, not enough. There are many conversations with senior leadership, and I, as chief attorney for the Coast Guard, get to be a part of those conversations. Probably more than some would like, and it occurs to me that some of these things are structural. When I hear them say, "We can't keep women in our service," that's because we design something like it's the 1950s when some of these "things" don't have to be done. It is so hard for two parents to work it out, and it seems like it's more women getting out than men. We must look at employment differently. And we need to make service in the Armed Forces a little more fluid, make some accommodations, and do actual talent management. When we talk about advancement, I see many people who are just ticking off all the boxes, but that's not realistic for anyone wanting to have a life with a spouse who is also interested in pursuing their own career.

-Melissa B., O-8, US Coast Guard
Chief Counsel and Judge Advocate General and Mother of one

I feel proud when I hear my son hum or sing along with the National Anthem. Knowing that he understands our flag's importance and our service makes me feel proud.

What kept me in? One was a move and the service obligation of the move, and the other was an educational program I enrolled in. I love the Marine Corps but saw that it was not sustainable for my husband and me as a dual military couple. And I am an "all or nothing" sort of person, and I didn't think the reserves were for me. But because I love the Marines, I decided to try the Reserves route to see if that's for me. The Reserves give me the support and benefits that the active-duty personnel get, and I really like the continued support that the Reserves provide.

-Janell H., O-3, USMC Reserves
Logistics Officer and Mother of one

Being a mom has definitely helped me be a better leader. I also feel like the shortcomings I felt as a mom, I can now use to mentor and help other women going through the same thing. Sometimes it is just about sharing a story or letting them know how it works. Since becoming a mom, my ability to relate has been so much better. As a military brat, I thought I could relate to others, but now the depth of relationships I can develop is more substantial. It's been really cool, especially with my new role in Civil Affairs.

I continue my service because I like it and want to be there. Passionate is a great word to describe my feelings about the Marine Corps; I thoroughly enjoy the people and the mission. It breaks my brain to think about not serving.

-Corey G., O-3, USMC Reserves
Combat Engineer Officer and Mother of two

I think a lot of people believe what I do is pretty cool, and it's a great feeling. It's rewarding to know that you are being recognized for your efforts and can juggle the two areas (service and motherhood). To know that others who understand the challenges, even my senior leadership, are incredibly impressed. Probably because they don't see the stress it takes to do it sometimes. It's rewarding and very flattering, but I think sometimes there is a negative connotation to what I do. And feel that some think there is a piece of me that is not wholly committed to the Marine Corps, and it shouldn't be like that.

What keeps me in? I would say a lot of things. My family because I have a responsibility to support them. But also, my passion for what I do is always a reason. It's kind of split; I joined the service to serve my country and defend our nation. That is still there, but my priorities have changed; I also have a family. As you become a leader, the idea of dedication twenty-four/seven now has a different meaning in every aspect - as a mother, a wife, a Marine Corps officer, and a Christian. There are multiple levels of responsibility and dedication to all of these aspects.

-Maria F., O-3, USMC
Training Officer and Mother of two

A couple of assignments back, I finished a very demanding assignment. And when I left, the going away gift I received was a wonder woman figurine dressed as a Marine. It meant so much that my superiors and peers recognized all my hard work. It was super sweet.

I try to be the leader I didn't have coming up in the Marines. I want to be the advocate for young female Marines who don't know where to turn for assistance.

I started a Women's Leadership and Education forum where all women (lowest to highest ranks, officer and enlisted) can get together and talk about issues that impact us because I wish I had that same opportunity for myself early on. I think that is one of the biggest rewards for me.

What keeps me in? I enjoy serving, and at the rank of E-8, I can take a step back, delegate more, and mentor more. I can focus on, listen to their issues, and help them become better marines and better people. That's what I really like.

-Jennifer P., E-8, USMC
Administrative Specialist and Mother of two

I don't know if there are rewards to being a mom while serving, but I get a lot of personal satisfaction in serving. I get to do incredible things. I think I am making a difference in the world, in America, and in the individual lives I touch and am responsible for. I don't know if my being a mom or not makes an impact. Maybe I am a better leader; I come with more empathy and understanding because I do have children. I can better understand other people, their situations, and maybe even the emotional roller coasters people are on. I don't know; perhaps, in some ways, being a mom has helped me be a better person at work.

What keeps me in? I would say good leaders, good bosses. I sometimes joke with people that I have been getting out since I got in. But really, I would sit down with my bosses and address everything and just end up staying in. At some point, you get to the point where you are close enough to twenty years, so you stay. But it started with male bosses who didn't understand why I would want to get out and would work with me and encourage me to stay in. Now, I need to stay in a bit longer to retire at this rank.

I have had more opportunities in the Air Force than I would have on the outside. At different times, it was various reasons that have kept me in. I am not in a position where I don't have to work, and I know how to do this. And to be honest, as a Black woman, I need to have as much work experience and rank as possible when I retire to compete on the outside.

-Rojan R., O-6, USAF
Program Manager and Mother of three

CHAPTER 10: IMPACT OF MOTHERHOOD ON SERVICE

For me, being a mother made me a better professional, because coming home every night to my girls reminded me what I was working for. And being a professional made me a better mother, because by pursuing my dreams, I was modeling for my girls how to pursue their dreams.[17]

- Michelle Obama

Serving in the Armed Forces is more than a job; it permeates every part of your life. Because of the military's mission (defense of our nation), it must. It requires much of us if we are to protect and defend our country and our allies.

As such, it impacts every corner of your life: from where to live, to what you do, to where you go, to when and where you can take your family along. Our service to our country touches every part of our lives, from big to minor, including our roles as moms.

Being a mother can also permeate every part of your life. It is a responsibility that weighs on you daily and is never-ending. The responsibility for another life is an awesome one, awesome in its joy and the weight of obligation it carries. Not just to the child but to the adult the child will become. And even to the world our children will grow up and contribute to. Raising a child requires a great deal of dedication and motivation. It requires you to be a mother, even when you don't feel or know if you have the strength to do it.

Inherently, motherhood is your legacy to your children and their children. It is also your legacy to the world. As moms, we make many decisions while our child is growing up. Decisions that directly and indirectly impact them. These decisions are weighed against our roles as moms and sometimes our roles as servicewomen.

For many of us, our desire and goal are to serve our country well and to raise our children to be productive and well-adjusted citizens. The goal is to let our two worlds (service and motherhood) intertwine without one choking out the other, to effectively serve in the Armed Forces and successfully raise your children. Both require much of us and, in a lot of ways, influence the other, whether we like it or not. And, for most of us, the impact is positive, both on our service and our children.

Sometimes, our role as servicewomen leaks into our role as mothers and vice versa. Because both are so important, they tend to bleed across the line that separates them. As a result, we find ourselves bringing our service into our role as moms and bringing our roles as moms into our service. This crossing between the two is not a bad thing. You can often take the discipline and order you gain through service and apply it to your role as a mother and take the patience and empathy you gain through motherhood and apply it to your role as a servicewoman.

While interviewing my fellow servicewomen, more than one related that the skills learned as a mother helped them become better leaders to their Soldiers, Sailors, Marines, Airmen, and Coast Guardsmen. Admiral Melissa B. said it best in my interview with her when she related leadership to motherhood, pointing out that the qualities that make you a successful leader mirror the qualities that make you a caring mother. As I talked with these women, I realized, even more so, that motherhood is not detrimental but complementary to serving in the Armed Forces.

> *I became a mother so early in my career that I can't define my service without including my role as a mother. Becoming a mother gave me a greater sense of responsibility, which flowed into my role as a servicewoman. Being a mom helped me think and make decisions that impacted my child and me; therefore, it was easier to think about and make decisions as an airman and later a supervisor and leader by considering the bigger picture. Early on, I learned to think of others and make decisions based on the benefits for all, not just myself. Because I grew as a mom and servicewoman simultaneously, it was easy to see that these roles were intertwined and influenced by one another.*
>
> -Patricia Q., E-7, USAF (Ret)
> First Sergeant and Mother of three

> *My role as a mother has impacted me and allowed me to see things from other perspectives. When I was younger, I was really by the book. But once I had children, it helped me to step back and see things from other people's perspectives, especially that of other parents. It helped me become more well-rounded and see and understand how our lives can affect others and how our lives can affect our jobs.*
>
> -Anna P., E-5, USAF
> Services Specialist and Mother of two

> *My role as a mom impacted my service, but only in the eyes of those who didn't understand what it is to be a mother. For example, I remember one of my commanders (who had a stay-at-home wife) was very insensitive to my role as a mother. During a regular meeting, my daughter's school called to tell me she was sick and that I would need to pick her up. Rather than let me leave, he wanted to know why I could not activate my family care plan, which was a plan that we only used for deployments and TDYs.*

Often, I had to stay up late to take care of stuff that others might be doing right after work, but I would use that time to spend with my daughter.
-Pagerine J., O-5, USAF (Ret)
Medical Service Corps and Mother of one

Has there been an impact? Yes, in some instances. For example, as a mid-grade Captain, I had just had my fifth or sixth child, and my commander basically asked me if I could keep up because of the number of children I had. Also, I know that some of my officer evaluations were not what they should have been because of my family dynamics (the number of children I have). And then, when my twenty-five-year-old son was in a near-fatal accident, we were trying to decide whether to take him off the ventilator (he is good and a walking miracle). My boss, at that time, gave me a write-up that did not match my performance. Because I put my short-term priority on my son, he tried to let my annual evaluation reflect that short period and negate my performance for the previous eleven months. Thank God my next boss realized that the previous one had not set me up for success and that my status during this time should have been coded in the system as "not available" instead of holding my situation against me. I certainly have had my share of people trying to use the fact that I have twelve children against me. But I do my job well.

That is another reason I am encouraged to stay on active duty. If or when I encounter those situations, I will address them and make it clear that this is not the atmosphere the Army needs to have in terms of women in uniform who are mothers.
-Leslie L., O-5, USA
Human Resource Officer and Mother of twelve

For me, being a mom gave me a greater level of patience. I remember teaching a young group of officers, and being a mom gave me a greater sense of tolerance and greater respect for authority. Having to make the tough decisions and do what's right even if those you lead (or raise) don't understand.
-Yulanda B., O-4, USAF (Ret)
Personnel Officer and Mother of one

Not in the way most people would think. I worked with a lot of soldiers and their families. And I think being a mom made me more open and allowed me to have a softer side. I still got on the soldiers when needed, but I also had more empathy. I remember one soldier I was working with, and I had to tell him that he was screwing up, but I still had to be able to help him with his situation. Being a mom made me a little more mindful, especially with single moms trying to do both roles by themselves. Many of them are working hard and trying to make both roles work. So being a mom changed my outlook a little bit.

-Eboni B., E-8, USA (Ret)
Career Counselor and Mother of two

It has definitely had an impact. The Army JAG Corps' unwritten rule is that only one can succeed when it comes to dual military couples. They felt that it was either stay together geographically or succeed while separated. So, without even trying, they just decided to take that stance. I had a female boss, who went on to be one of the highest-ranking generals in the Army, have an issue with me because I didn't choose to have an au pair or nanny but instead decided to raise my children myself.

I personally know of eight African American women in the JAG Corps eligible for promotion between 2011 and 2015. All of them except one was a mom, and zero got promoted; they had excellent appraisals, but because they were parents had been given jobs that marginalized them and reduced their competitiveness.

-Patricia H., O- 5, USA (Ret)
Judge Advocate and Mother of three

Becoming a mom has helped me be a better leader. Before becoming a mom, I was harsh and didn't have the patience to deal with issues outside the military realm. Like many people I worked with and worked for, I was hard-charging and not really patient.

Being a mom has helped me be more empathetic as a leader. It has helped me be a better friend and colleague. It has helped me better understand and be able to advise other parents without being judgmental.

Most of the leaders that I know who are also parents are better leaders. I didn't realize that until after I had children.

-Toni D., E-8, USAF
Personnel Specialist and Mother of three

I felt there were higher expectations of me as a black woman and a single mom, and I really felt like I had to work harder to compete. I think some people don't get the recognition they deserve, and I also fell into that group. As a soldier, I felt pushed harder because of who I was; a Black woman and soldier.

-Kimberly B., E-5, USA Veteran
Logistics Technician and Mother of four

Thinking about it, my role as a mom has definitely impacted my service. Even at this stage (still in the first years), coming on active duty as an older person and a mom, I think I have been able to see the benefits and help encourage younger troops. Mentoring younger airmen and being a mom have helped me "be a mom" to the younger airmen.

-Jennifer H., E-3, USAF
Intelligence/Cyber Specialist and Mother of one

All the impact has been positive. I think I have gained so many new tools as a mother that I have been able to transfer to work. I don't feel that I that there have been any negative effects on my career since becoming a mother. But as I said earlier, I was a bit older and established at work and that prepared me to better deal with work and motherhood. I came back to work with a greater sense of calmness and better able to manage my day-to-day work.

-Chelsea D., E-7, USAF
Medical Lab Technician and Mother of three

It has definitely impacted my career. Being a mother has taught me to think before I do; sometimes, in the Navy, they expect you to do and not necessarily think about what you are being told to do. That can be problematic. Motherhood has helped me learn not to take some things so seriously and to just focus on the tasks, on what's important. My age, life experiences, and motherhood have helped me help those younger sailors coming in. My role as a mom has helped me be a "mom" figure to many of the younger troops.

-Mykeia L., E-3, USN
Logistics Specialist and Mother of five

With only one year in service, I can already see it. I also see that some people expect you to pick or choose service or motherhood. But I do not know why you wouldn't stay in the military.

And I think it's different when one's spouse is also military. Some people ask me how I do this with kids, and I think I do it the same way everyone else does with kids.

-Kelsey Q., O-2, USAF
Operating Room Nurse and Mother of two

I think, honestly, it made me stronger. It made me look at life differently. I knew I had to go to work and do the shifts. And when it came to making the hard decisions, I could sit down with my husband and make those decisions. And when my children came along, I looked more to the future and being able to provide for my family.

As a soldier, the military has made me more structured; and as a mother, the military made me look at life more seriously because you will experience things in the military that you won't encounter anywhere else. And you can't always call mom because they don't really understand what you do. So you have to figure it out.

-Cynthia S., O-4, USA (Ret)
Nurse and Mother of two

I did have to make some sacrifices. I feel like I was given some choices to make to see how far I would go, and there was only so far I was willing to go. And at this point, I am still willing to do what I can, but there were decisions I might not have made if I were single. But I am much happier now.

I remember being single and thinking about why people would bring their kids to work sometimes. And now, here I am doing the same sort of thing. It does change your perspective and being a mom has made me a much better leader and supervisor. It's what leaders do, quite frankly. Leadership has been labeled as paternal, but it's very maternal; it is what women do. So, women are great leaders naturally, I think.

I get some things so well now that I did not understand as a single person. Just the kinds of worries that parents have. I never understood many things, like moving around and trying to get daycare when there is a two-year waiting list. I wasn't tuned into that before, but I am very attuned to it now.

-Melissa B., O-8, US Coast Guard
Chief Counsel and Judge Advocate General and Mother of one

I think there are a couple of aspects of impact. I think the negative side of it is easier to say out loud. I can't put in as many hours, and I may not necessarily live up to the image of what comes to mind for most people when they think about a Marine officer. But when I put all that aside, what I did bring to the table was completely different. And so, I believe that being a mom in the military helped me be a better leader, more patient and understanding of those around me, and how to maximize my time. Because I didn't have time to waste at work, I had to be much more focused. If I am there to work, I need to focus on that and get it done so I can go home and focus on home.

-Janell H., O-3, USMC Reserves
Logistics Officer and Mother of one

I don't necessarily wish I had children sooner, but being a mom made me a better person and leader. It impacted my worldview and perspective. When I look back, there were times when I could have made a better decision with folks who worked for me. And when I do look back, I am like, "Did I really say that?"

I became more patient, which was better for everyone around me. I was able to make decisions patiently. Some of that comes from experience, but I feel it was more evident for me after I became a mom. I don't get upset as much; I can really put things into perspective. I have a stronger sense of calmness.

Maybe from a negative point, I thought I was ready to return to work. In my mind, after having my child, I just wanted a more regular schedule that work brings. But, three months after giving birth, my brain was re-wired, and it was hard to catch up and process things. I didn't enjoy that part. So, I am glad they have extended servicewomen's recovery timeline after giving birth.

-Corey G., O-3, USMC Reserves
Combat Engineer Officer and Mother of two

I definitely think my role as a mother has positively impacted my service and my role as a leader. It has made me a more compassionate leader, which is an essential aspect of leadership. I don't see a lot of that quality of leadership in leaders, but I think that my marines value that in me. I have more compassion for the warfighters I lead, and I think they appreciate that I can sit down and listen to what they have to say first. I believe that leaders make leaders. If you can start with understanding, listening, and being compassionate in what they are dealing with, that is one step closer to building a more cohesive environment.

-Maria F., O-3, USMC
Training Officer and Mother of two

I remember being an E-6, which is middle management, and being pregnant, I had to go to the rifle range, and they told me that it was a biohazard because I was lactating and would leak all over the weapons. And if they were telling me this as an E-6, what were they telling our younger marines? It was crazy.

Taking time out to be a mom, I chose to breastfeed, and I would have to take time out to pump in the morning and then go to the Childcare Center to feed my child. So that was more time away, you know. And not every unit makes accommodation for that. Even in my last unit, I had to pump in the bathroom.

Being a mom has changed my outlook on how I view and treat marines. I find myself being the "office mom," making sure everyone has what they need. Being a mom has made me a better person and a better Marine. My folks trust me and know that I have their best interest in mind, that I will keep what they tell me in confidence, and I will do what I can to help them. And now that I have kids, I can relate to and help junior marines who have young kids. The Marine Corps is very male-dominated, and I feel comfortable saying that many of them have stay-at-home wives, so they don't deal with many of the things that moms must.

-Jennifer P., E-8, USMC
Administrative Specialist and Mother of two

121

Being a mom has definitely impacted my career, 100%. But I am different, and I am very vocal about this. And now I am in this weird position where they cannot fire me. I can go a little rogue. I have authored articles, testified to the House Committee, and written a book. I am trying to vocalize to all the men I work with to talk about what you cannot say or shouldn't do, things like doing PT at 6:00 am when the childcare center doesn't open until 6:30 am. To help them look at things from a different perspective.

I want to be a voice to try and improve it and uplift people so they can fulfill their dreams.

So yes, being a mom has totally helped me, and maybe I am becoming who I am supposed to be through this whole process. Like being more vocal, sitting at the table with congressional people, and telling them what it's like as a mom in the service. There are some excellent advocates on the Hill right now, and I appreciate being able to tell them what our world is like; I say those things they need to hear for both women and men.

It's not all roses. There are still people out there who are jerks and won't be supportive. And I tell women, you have to decide what you will do in that situation. Will you allow that one person to take you off the course to your dream, your goal, or what you need for your family? Or are you going to find a way so that this one person can't take you off the path you set for yourself? There will be jerks, obstacles, those people who are not possibility-minded for you, or whatever you want to call them. You must decide what your reaction will be in that situation. It is hard, but I always believe that it is not hard forever. I mean, there were whole assignments that were hard, but I look back on it now, and it was just a blink. Those years go by, and you forget about them because they are surrounded by good years. It's in those struggles that you must decide what you are going to do. You keep getting up, showing up, and then you are in a good place.

-Rojan R., O-6, USAF
Program Manager and Mother of three

CHAPTER 11: ONE LAST THOUGHT

"I am a Woman Phenomenally.
Phenomenal Woman, that's me."[18]
- Maya Angelou

While serving in the Armed Forces requires much of us, we gain as much, if not more, from our service. There is a lot to be said about serving in uniform. Some are good, some not so good, and sometimes it is hard to put into a few words what we want to share with those who may or may not understand the experience or the decision to serve while raising our children.

Serving in the Air Force was absolutely a wonderful experience for me. Looking back, I can see that I gained so much and grew so much as an airman, NCO and a person. When I look at my children, I can see their lives and how their experiences as military brats helped to shape them into the man and women they are today. When I look at the future, I see the women who have stepped up to the challenge of serving in uniform, and I am hopeful that their experiences and insights in this book will prove helpful to other women who choose children and careers.

On more than one occasion, the women I spoke with helped solidify my decision to write this book. I hope this book is a source of encouragement for other moms in the military, servicewomen who decide to stay in uniform after becoming a mom, or civilian moms considering joining the Armed Forces. Unfortunately, there is no instruction book for being a mom and no regulation for being a mom in the military.

While we all do the same two things: service and motherhood, we do it differently, for different reasons, and even cope in different ways, but we all do it. Whether we did in the Vietnam era, the Cold War era, the Desert Storm era, or today, we served in the Armed Forces and raised our children.

I wrote this book to help other moms understand that they are not alone and provide some helpful insight into their roles as servicewomen and mothers. I also hope this book will encourage and help them see that they can do it. Knowing that the woman before you did it can be encouraging, and I hope the stories of the women in this book encourage other women. To help them realize that they may have to work a little harder, find workarounds, find their own source of support, and study a bit longer to get what they want, but service and motherhood are possible. I hope they can see themselves as strong and capable women able to tackle the roles of servicewoman and mom.

I hope this book provides insight into why we chose to raise our children while serving in the Armed Forces. This book could help a family member, a fellow serviceman or woman, a supervisor, a friend, a neighbor, or even someone who has wondered why we do what we do.

I could not think of a better way to end this book than with final thoughts from the women included in it, and to each of them, I say thank you for your service and for being a part of this project.

You may never know the positive impact of your decision on others. You never know how many people are watching you and will follow in your footsteps. When I joined the Navy, there were not a lot of women in the service, especially not a lot of women of color. However, I have found that many women I know followed suit after seeing me join the military. I am proud to have had such an impact.

-Jacqueline G., E-3, USN Veteran
Personnel Specialist and Mother of two

The military offers its members many benefits, including the opportunity to travel and meet new people. It also has benefits like paid vacation, medical care for you and your family, and full paid maternity leave. But more than that, the military is the opportunity to serve your country and, at the same time, teach your kids, especially as they get older, about service and the sacrifices you are making for your country.

-Anna P., E-5, USAF
Services Specialist and Mother of two

You never know how much you will have to sacrifice while serving in the military, but there are rewards. Our active-duty members make so many sacrifices through their service, and it is a whole different lifestyle.

-Carol G., USAF Spouse
Teacher and Mother of one

I think we have to remember to put things in perspective. Sometimes you will feel bad or guilty about your service, but you have to put things in perspective and look at the big picture. A friend once reminded me that my child would not remember that one time when I had to work rather than play on the playground with her. Keeping that in mind helps.

-Pagerine J., O-5, USAF (Ret)
Medical Service Corps and Mother of one

I think it's important that you shape or rewrite the narrative that you can be a mom and be a soldier, Airmen, or Sailor, and you can do both well. For me, it's making expectations clear and setting boundaries and parameters with your leadership. That is what I have done for almost twenty-three years. I will give everything I have to the military and the mission, but I will not sacrifice my family, and it has worked for me.

And don't be afraid to tap into the resources. As long as I have been in the service, there have always been many different resources. Even as a commander, I found that many of our troops don't even know about the available resources. As leaders, we need to lead our people to these resources. Because if you are no good at home, you are no good to me.

-Leslie L., O-5, USA
Human Resource Officer and Mother of twelve

I remember one of my commanders calling me into his office. He explained that when I walked into a room, my presence would be felt by those in the room. Simply because I was a female and a woman of color, which I already knew, and he explained that I would have to work harder than my male counterparts and they would not receive as much pushback as I would. But I was always determined to do what I set out to do, even with the challenges.

You know, people often ask me how I do it. The truth is you just do it. You do what you have to do. Just as women have done for so many years and even decades.

-Yulanda B., O-4, USAF (Ret)
Personnel Officer and Mother of one

126

It's not fair. But as a woman, you must be that much better than any man because you have a stigma attached to you (as a woman). Unfortunately, I feel like that is where we are as women today.

It's a struggle sometimes, but there will always be another woman who can see you and your struggles and help you out. That is how it happens, and that is how we give back.

-Eboni B., E-8, USA (Ret)
Career Counselor and Mother of two

Every person must decide the path they will take. My advice is not to let others determine your value, worth, or path in life.

-Patricia H., O-5, USA (Ret)
Judge Advocate and Mother of three

As a senior NCO, I know there is a stigma attached to being a mom in the military. To this day, it is still happening. I remember a senior leader asking me if I thought women were not scoring high enough on promotion testing because they were too busy taking care of things at home. It was clear to me that is a stigma that women have too many responsibilities at home to be competitive. So, it's rewarding for me to be able to show others that women who are moms can serve and succeed in the military.

-Toni D., E-8, USAF
Personnel Specialist and Mother of three

You don't get a handbook; you just do the best you can. We have always tried to allow our children to spend as much time as possible with extended family. We made it a point to spend time with family, especially when I was not working. You know, summer and spring break were an excellent time to allow them to spend time with and get to know their family.

-Ebony A., USAF Spouse
Mother of three

The military allowed me to grow up faster, and I gained a sense of discipline. For me, in the end, there were too many negatives. It's essential to know yourself, what you can give, and what you need.

-Kimberly B., E-5, USA Veteran
Logistics Technician and Mother of four

I hope this book helps other women interested in joining the Armed Forces. But unfortunately, I think there is still a stigma, and many women might be hesitant to join; they may not have other women around them to answer their questions about life in the Armed Force.

-Jennifer H., E-3, USAF
Intelligence/Cyber Specialist and Mother of one

I love being in the Air Force and being a mom; I have the opportunity to do two things that make me happy. When I have had a tough day, it's nice to go home and see my children; they keep me grounded and make up for anything that happens at work.

-Chelsea D., E-7, USAF
Medical Lab Technician and Mother of three

Even when it seems too much, I know that my service is the best decision for my family and me; and I am showing my children that whatever you want to do, you can do. I know that if my children find themselves in a situation that seems too hard, they can look back at me and my service and know that if their mom made it happen with five kids, they could do it as well. I want my children to know that anything is possible.

-MyKeia L., E-3, USN
Logistics Specialist and Mother of five

The military makes you a stronger person. You discover who you are and how strong you are as a person and as a family unit. You grow and see how stable your relationship is and how much you really lean on each other. You learn that your family bond is strong and can't do it without them. I think many people assume that just because you are in the military, you are always going to be away from your family; that is simply not always the case. And even if you are not in the military, many civilian positions still require you to travel frequently for work.

-Kelsey Q., O-2, USAF
Operating Room Nurse and Mother of two

I can't wrap my brain around women being unable to serve as mothers. So many great things have come from the women serving. Like in the civilian sector, the Nurse Corps is predominately female, and they have contributed so much to the Armed Forces. I have served with a plethora of great women who were also mothers. In service, just like in motherhood, we make things happen.

And in the military, unlike the civilian sector, you get so much from working together the way we do. The ability to work together under a variety of conditions and camaraderie is something we develop from the beginning, and that helps you in all aspects.

-Cynthia S., O-4, USA (Ret)
Nurse and Mother of two

I think this book is a great message and is not said enough. I have even had senior women say, "I couldn't be that kind of a mom, so I decided it was career or motherhood," and I thought, is that really what we are telling our young women? However, I have yet to meet a senior male officer who says it was his career or fatherhood.

There will be times when you drop the ball, and it happens to all of us, men and women. You might go through periods when you have other priorities but don't walk away if you don't want to. We, as women, are so hard on ourselves, and sometimes it is okay to not be at the top and not beat yourself up over it. You keep going.

-Melissa B., O-8, US Coast Guard
Chief Counsel and Judge Advocate General and Mother of one

No one can do it all, no matter how hard we try. You must be realistic and establish a sense of balance. For me, because I wanted to continue serving, the Reserves became my path. It's essential to find what works for you. Don't be afraid to try different things and find your path.

-Janell H., O-3, USMC Reserves
Logistics Officer and Mother of one

It is encouraging to see the changes that are happening for moms in service. I think there are so many good things right now, and I hope it keeps improving for women.

I got to attend a joint-service women's event while I was six months pregnant, and I was so glad to be a part of that. A civilian government employee from the Manpower Branch responsible for writing the upcoming changes approached me, and I was excited to share my thoughts with him. During this conference, I learned that women from every branch, much more senior than I, are doing service and motherhood.

And for any servicewoman getting ready to experience motherhood, you have your entire pregnancy to get things in order at work. So you should ensure leadership is helping you plan for your absence rather than acting like your absence is so unexpected when the time does come.
-Corey G., O-3, USMC Reserves
Combat Engineer Officer and Mother of two

I don't want to sugarcoat it. It has been challenging; my husband has given up his position to allow me to take on a great opportunity within the Marine Corps and stay together as a family unit. We made a considerable sacrifice for me to do that. But we've adjusted, and our faith has sustained us.
-Maria F., O-3, USMC
Training Officer and Mother of two

I like the possibilities and opportunities that the Marine Corps has given me. I am E-8 now and have been grinding for almost twenty years, and I still like what I do. I like that I can now take a step back, give back, and mentor my troops more. I want to be able to make them better Marines and better people.
-Jennifer P., E-8, USMC
Administrative Specialist and Mother of two

Our careers and lives are not structured the same way as everyone else. I like to remind people that the things we do to keep our families together and our careers together don't have to be like everyone else's because our careers (in the military) are so different.

So, I encourage people to get creative in their solutions and not look at the person to the left or right and think that what they do will work for you, because it may not. You may have to let go of a lot of money or re-think some things to keep your career going and your kids thriving in the things they want to do. But you can do it.

And it's okay to cry; you will not be the only woman to cry when you cry. We all cry, and some of it is not even us; we have a lot going on inside of us, and our tears are sometimes how we deal with things. You have to find a radius in which you can operate, be comfortable and successful, and get all the things you and your kids need. Is everything close enough for you to operate within that radius? Things like driving distances, a good school, before and after school care, and extracurricular activities. That's what we do.

-Rojan R., O-4, USAF
Administrative Specialist and Mother of three

APPENDIX 1
Rank Structure

To provide a broad view of the responsibilities that come with each rank, included in the appendix are excerpts from the DoD about each level of ranks and responsibilities[19.]

Service members in paygrades E-1 through E-3 are usually either in training status or on their initial assignment. The training includes the basic training phase where recruits are immersed in military culture and values and are taught the core skills required by their service component.

Leadership responsibility significantly increases in the mid-level enlisted ranks. This responsibility is given formal recognition by use of the terms noncommissioned officer and petty officer. An Army sergeant, an Air Force staff sergeant and a Marine corporal are considered NCO ranks. The Navy NCO equivalent, petty officer, is achieved at the rank of petty officer third class.

Senior Noncommissioned Officers include the upper rank of the enlisted structure. At the E-8 level, the Army, Marine Corps and Air Force have two positions at the same paygrade. Whether one is, for example, a senior master sergeant or a first sergeant in the Air Force depends on the person's job. The same is true for the positions at the E-9 level. Marine Corps master gunnery sergeants and sergeants major receive the same pay but have different responsibilities. All told, E-7s, E-8s and E-9s have fifteen to thirty years on job are commanders' senior advisers for enlisted matters.

Warrant officers hold warrants from their service secretary and are specialists and experts in certain military technologies or capabilities. The lowest-ranking warrant officers serve under a warrant, but the receive commissions from the president upon promotion to chief warrant officer 2. These commissioned warrant officers are direct representatives of the president of the United States. They derive their authority from the same source as commissioned officers but remain specialists, in contrast to commissioned officers, who are generalists. There are no warrant officers in the Air Force.

The commissioned ranks are the highest in the military. These presidential commissions and are confirmed at their ranks by the Senate. Army, Air Force and Marine Corps officers are called company grade officers in the paygrades of O-1 to O-3, field grade officers in paygrades O-4 to O-6 and general officers in paygrades O-7 and higher. The equivalent officer groupings in the Navy are called junior grade, mid-grade, and flag.

APPENDIX 2: (ENDNOTES)

[1]Hegar, Mary. Women Warriors Are On The Battlefield, Unfair Military Combat Exclusion Policy, ACLU.org, https://www.aclu.org/news/womens-rights/women-warriors-are-battlefield-eliminate-outdated-unfair

[2]Truman and Women's Rights. (n.d.). Truman Library Institute. Retrieved August 15, 2022, from https://www.trumanlibraryinstitute.org/truman-and-womens-rights/#:%7E:text=Signed%20on%20June%2012%2C%201948,four%20branches%20of%20the%20military.

[3]EXECUTIVE ORDER 10240 | Harry S. Truman. (n.d.). National Archives. Retrieved August 11, 2022, from https://www.trumanlibrary.gov/library/executive-orders/10240/executive-order-10240

[4]Military Women and Pregnancy. (n.d.). Userpages.Aug.Com. Retrieved August 11, 2022, from http://userpages.aug.com/captbarb/pregnancy.html#:%7E:text=In%201976%20the%202d%20District,was%20permanently%20unfit%20for%20duty

[5]Murnane, L. S. M. (2007). Legal Impediments to Service: Women and the Rule of Law. Duke Journal of Gender Law and Policy, 14(2), 1061–1097. https://scholarship.law.duke.edu/djglp/vol14/iss2/

[6]Levine, K. (Writer), Tyne, G. (Director). (1978, Jan 30). What's Up, Doc? (Season 6, Episode 19) [TV Series episode]. Metcalfe, B. (Producer). M*A*S*H. 20th Century Fox Television

[7]2020 DEMOGRAPHICS PROFILE ACTIVE DUTY MEMBERS. (n.d.). [PDF]. 2020 Demographics Profile of the Military Community. https://download.militaryonesource.mil/12038/MOS/Reports/2020-demographics-report.pdf

[8]Demographics of the U.S. Military. (n.d.). Council on Foreign Relations. Retrieved August 11, 2022, from https://www.cfr.org/backgrounder/demographics-us-military

[9]House: Banning Military Mothers Was Wrong. July 15, 2022, Associated Press of the United States Army. Retrieved October 29, 2022. Website: https://www.ausa.org/news/house-banning-military-mothers-was-wrong

[10]Amelia Earhart Quotes. (n.d.). BrainyQuote.com. Retrieved Aug 24, 2022, from BrainyQuote.com Website: https://www.brainyquote.com/quotes/amelia_earhart_163002

[11]Rose Kennedy Quotes. (n.d.). BrainyQuote.com. Retrieved October 29, 2022, from BrainyQuote.com Web site: https://www.brainyquote.com/quotes/rose_kennedy_597708

[12]Ayn Rand Quotes. (n.d.). BrainyQuote.com. Retrieved October 29, 2022, from BrainyQuote.com Website: https://www.brainyquote.com/quotes/ayn_rand_105316

[13]Epicurus Quotes. (n.d.). BrainyQuote.com. Retrieved October 29, 2022, from BrainyQuote.com Website: https://www.brainyquote.com/quotes/epicurus_161673

[14]Sophia Loren Quotes. (n.d.). BrainyQuote.com. Retrieved October 29, 2022, from BrainyQuote.com Web site: https://www.brainyquote.com/quotes/sophia_loren_106851

[15]Chanakya Quotes. (n.d.). BrainyQuote.com. Retrieved October 29, 2022, from BrainyQuote.com Web site: https://www.brainyquote.com/quotes/chanakya_201070

[16]Beyonce Knowles Quote. (n.d.). From QuoteFancy.com. Retrieved October 29, 2022 from Fancy Quote website: https://quotefancy.com/quote/1277660/Beyonc-Knowles-We-need-to-reshape-our-own-perception-of-how-we-view-ourselves-We-have-to

[17]Michelle Obama. (n.d.). AZQuotes.com. Retrieved October 29, 2022, from AZQuotes.com Web site: https://www.azquotes.com/quote/1463178

[18]Maya Angelou, "Phenomenal Woman" from And Still I Rise.

[19]Military Rank Insignia. (n.d.). From U.S. Government.com. Retrieved October 29, 2022 from US Government website: https://www.defense.gov/Resources/Insignia/

ABOUT THE AUTHOR

PATRICIA QAIYYIM was born in Michigan City, Indiana. She grew up in a large family and lived in several states throughout the Midwest. From life in small-town Indiana to farm life in rural Mississippi, with a few states in between. While in high school, Patricia spent time in the Junior Reserve Officer Training Corps (JROTC). After high school, Patricia attended college for two years before joining the States Air Force. She served twenty-three years on active duty. She has three siblings who also served in the Armed Forces.

As a member of the Air Force, Patricia had the opportunity to work in several specialties (jobs) as an Aerospace Ground Equipment Technician, a Military Training Manager, a Paralegal, and a First Sergeant. After basic training at Lackland Air Force Base in Texas and her first technical training at Chanute Air Force Base in Illinois, Patricia's first assignment was Zaragoza Air Base, Spain, where she met her husband. Her other assignments included installations in Texas, Arizona, South Carolina, and Japan. She has traveled to many parts of the United States and several countries during her service to her country while raising her three children. Patricia enjoys learning about other cultures, making all her assignments an adventure and learning experience. She has always tried to pass that love on to her children.

Patricia became a mom early in her military career and spent most of her career tackling service and motherhood. Her husband was also a member of the United States Air Force. Together, they served more than forty-five years in the Air Force. After retirement, Patricia continued to be affiliated with the Air Force as a contractor for more than seven years.

Patricia has always enjoyed a love of reading and writing. After many years of thinking about writing a book about life as a mom in the military, she completed her first book. She is currently working on a book of poetry and a companion book to Moms In The Military.

Besides reading and writing, Patricia is an avid quilter and enjoys everything crafty. She also enjoys cooking, working with her hands, and spending time with her family. She considers herself a woman of faith, a true Renaissance, and a citizen of the world. She believes we should all strive to inspire, motivate, and educate others as we move through life.
Patricia and her husband have three children and two grandchildren and currently live in Texas.